The Evolution of the Eye

Georg Glaeser · Hannes F. Paulus

The Evolution of the Eye

Georg Glaeser
Universität für Angewandte Kunst Wien
Wien, Austria

Hannes F. Paulus
Department für Integrative Zoologie
Universität Wien
Wien, Austria

Translation from German language edition:
Die Evolution des Auges - Ein Fotoshooting
by Georg Glaeser and Hannes F. Paulus
© Springer Spektrum/Springer-Verlag Berlin Heidelberg 2014
Springer Spektrum/Springer-Verlag Berlin Heidelberg is a part of Springer Science+Business Media

ISBN 978-3-319-17475-4   ISBN 978-3-319-17476-1 (eBook)
DOI 10.1007/978-3-319-17476-1

Library of Congress Control Number: 2015949439

Springer Cham Heidelberg New York Dordrecht London
© Springer International Publishing Switzerland 2015
This work is subject to copyright. All rights are reserved by the Publisher, whether the whole or part of the material is concerned, specifically the rights of translation, reprinting, reuse of illustrations, recitation, broadcasting, reproduction on microfilms or in any other physical way, and transmission or information storage and retrieval, electronic adaptation, computer software, or by similar or dissimilar methodology now known or hereafter developed.
The use of general descriptive names, registered names, trademarks, service marks, etc. in this publication does not imply, even in the absence of a specific statement, that such names are exempt from the relevant protective laws and regulations and therefore free for general use.
The publisher, the authors and the editors are safe to assume that the advice and information in this book are believed to be true and accurate at the date of publication. Neither the publisher nor the authors or the editors give a warranty, express or implied, with respect to the material contained herein or for any errors or omissions that may have been made.

Printed on acid-free paper

Springer International Publishing AG Switzerland is part of Springer Science+Business Media (www.springer.com)

The evolution of the eye

Tokay gecko *(Gekko gecko)*

VI

Seahorse *(Hippocampus reidi)*

Giraffe
(Giraffa camelopardalis)

Preface VII

While it may be impossible to photograph evolution per se, we may certainly photograph its results. Many famous products of evolution emerged independently across wildly different species, but nevertheless share functional or morphological features – often for very interesting reasons. Hence, a focus on specific organs or bodily parts is often especially useful, leading to additional, fascinating insights that a broader view would have missed. This book is a cooperation between an evolutionary biologist and a mathematician who also happens to be a passionate nature photographer. We aim to illuminate the vast field of evolutionary biology, wielding images and text with equal importance.

The majority of photographs were taken by *Georg Glaeser*, who aimed to depict the numerous animals with empathy and artistic grace, rather than with dispassionate scientific objectivity. The images on this spread are a good example of his approach – details emerge spontaneously at the expense of „zoological clarity", allowing even seasoned biologists to notice remarkable features or patterns of behaviour that would be missed on purely scientific images.

As a species, we rely disproportionately on our eyesight – a predilection that makes us prone to judging by appearance. Sea horses and giraffes may be extremely different animals, and yet, one would be hard-pressed to deny a superficial resemblance beyond the mere difference in size. Some even conclude, on the basis of appearance alone, that humans have fish-like ancestors – but looks may be deceiving, and our naive intuitions are often mistaken. Incidentally, evolutionary biology has already settled the question of what caused the giraffe's famous neck to grow so long: it did not evolve, as it is often believed, because the animals fed on trees that grew their foliage on increasingly higher branches in a desperate arms race to survive in the African savanna. Rather, their necks seem to have been molded by the ubiquitous mechanism of sexual selection.

In writing this book, we have attempted to give many examples along the various branches of the evolutionary tree. Our special focus always lies on the ways in which eyes have evolved. We have attempted to illuminate this gradual development from many perspectives, comparing the many strikingly different results of this billion-year process through

VIII

Tarsier *(Tarsius spec.)*

pictures and schematic illustrations. Since photographs play such an important part, the textual descriptions have been condensed to their essence. However, we have attempted to provide a broad scope of additional information through footnotes and related publications, including websites. This book is accompanied by a website of its own, where changes to the referenced websites are tracked and remedied whenever possible.

This book does not have to be read from front to back – in fact, it encourages a "coffee table" style of reading. However, we have tried to make jumping from chapter to chapter easier by including an extended glossary towards the end, which contains explanations of the most important terms, followed by an index for situations when even the glossary does not suffice.

We aim to present the staggering diversity of eyes in the animal kingdom – from stalk eyes to point eyes, facet eyes, and lens eyes. The simplest form of eyes already appears in single-celled organisms, which may possess a photoreceptor at the base of their flagellum, which allows them to distinguish dark from light. However, it is remarkable to notice that eyes are largely irrelevant to forms of life that do not possess a clear anterior (head, eyes, and feeding tools) and posterior (organs of excretion).

We would like to thank the following people for their valued cooperation on this book – in alphabetic order, omitting their academic titles: *Daniel Abed-Navandi*, *Gudrun Maxam*, *Axel Schmid*, *Manfred Walzl*, and *Sophie Zahalka*. Ms. *Stefanie Wolf* from Springer Spektrum Verlag has supervised the project with great engagement. Finally, we would like to thank *Peter Calvache* for his work on graphics design, layout, production, as well as for his translation from German to English.

Due to the limited number of pages available to us, we have had to select the printed photographs very carefully. Many interesting pictures didn't make the cut, but can still be found on this book's official website: www.uni-ak.ac.at/evolution

Western long-tailed hermit *(Phaethornis longirostris)*

Table of contents

Preface VII

1 Thoughts about evolution 1
Vertebrates and invertebrates 2
The theory of evolution –
A mathematician's point of view 6
A biologist's
point of view 10
The embryo and all stages
of the eye's evolution 14

2 Lens eyes or compound eyes? 17
The eyesight of articulate animals 22
Image raising through refraction 24
Two solutions
for sharp eyesight 26
Advantages and disadvantages
of the facet eye 28
The spatial vision of insects 30
Crab or insect? 32
Rolling eyeballs 34
Spider or crab? 36
Strange eyes 38
Facet eyes below the water 40
Newts and salamanders 42
Reptiles and amphibians
in the same biotope 44
The eyes of frogs 46
Morays and snake eels 48
The largest eyes
among land animals 50
Colorblind? 52
Cartilaginous fish 54
Birds and reptiles 56
Attacking the weak point 58

3 The world is 3D 61
The mantis and
the mantis shrimp 64
Precise estimations 66
The eyes of dragonflies
and damselflies 68
Life in the sand 70
Eight legs and a large brain 72

Harvestmen and octopuses 74
Double image processing 76

4 The limits of clarity 79
Telescope eyes 82
High resolution 84
Of crows and vultures 86
Round or slit-shaped pupils? 88
Low-light amplifiers 90
Shark vision 92
Two yellow spots
on the retina 94
Unbeatable 98
Are eight eyes better than two? 100

5 Simple or simplified? 103
A very primordial eye 108
"Pro forma" eyes 110
A diversity of miniature eyes 112
Where are the eyes? 114
Eight eyes do not make a
visually oriented animal 116
The simple eyes of insects 118
The square facets
of higher crabs 120
Fossils eyes also
deserve a closer look 121
Flying foxes 122

6 Above and below the water 125
A simple eye model
above and below the water 128
Adaptation to
vision under water 130
Short-sighted by nature 132
Fish at the surface 134
What do whales and hippos
have in common? 136
The eyes of aquatic mammals 138
How large
can or should eyes be? 140
Hot spots – targeted mutations 142

7 Pax and homology 145

8 Alternative senses 151
Smell as a replacement for sight 156
Sensing touch and vibration 158
Bats up close 160
An optimized sense of smell 162
Diurnal vipers and
nocturnal tree boas 164
Two orders of magnitude apart 166

9 A world of color 169
Stripes across the eyes 172
Quirky and venomous 174
Camouflaged eyes 176
The patterns on facet eyes 178
Pseudo-pupils 180

10 The language of our eyes 183
False eyes can be life-saving 186
False eyes under water 188
Intuitive oculesics 190
A comet's impact
65 million years ago 192
Close genes and close emotions 194
Eye language
of our closest cousins 196
Cuteness 198
Ambush predators 200
The hypnotizing eyes of vipers 202
Cuteness under water 204
Teddy bears or beasts of prey? 206
How many species are there? 208

Glossary 210
Literature selection 211
Index 212

XI

2 Lens eyes or compound eyes?

Two very different ways to see the world

The light-sensitive cells of all eyes — no matter their difference from ours — evolved either from the nervous system, or from cells of the skin. It only took those two types of cells to produce lens eyes and facet eyes, although only after a large number of generations.

1 Thoughts about evolution

Biological and mathematical ways of seeing

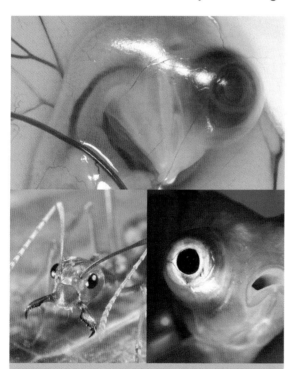

Biological evolution is the change in time through selection, which is the difference in reproductive success of a population due to varying fitness. This mechanism may noticeably alter the frequency of a trait over even a single generation.

3 The world is 3D

Spatial vision

The size and positioning of many animal's eyes reveal a lot about their lifestyles. Predator eyes tend to be directed towards the front, in order to capture their prey in three dimensions. The eyes of prey animals, on the other hand, are placed to maximize the field of vision.

4 The limits of clarity

Fine tuning of retina and optics

Any good eye requires high-quality optics and/or a high-quality retina. Birds of prey gain the most from improving their retina, while some crabs are remarkably successful at growing high-fidelity optics in each of their facets.

XIII

5 Simple or simplified?

When seeing clearly is not required

The simplest eye consists of just one sensory and one pigment cell. Another precursor of the lens eye – the pinhole camera eye – shows that "simple" does not always mean "primitive". Ancestors of millipedes had more complex eyes, which have become simplified.

6 Above and below the water

Different requirements for each medium

The eyes of water animals are different from those of their land-based cousins. Above the water, the outer cornea is more important to clear vision than the lens. Below the surface, the bulk of refraction is done by the spherical lens.

XIV

7 Pax and homology

The third act in the history of evolution

Genes for the development of light-sensitive cells can be found in the first single-celled organisms. From such early forms, the animal kingdom has evolved many types of eyes, all of which make use of common genetic functionality during embryonic development.

8 Alternative senses

When eyes are not enough

Animals inhabit a sensory world that mirrors the challenges that life puts before them. Apart from vision, the senses of smell, taste, sound, feeling, and temperature are ubiquitous among animals. Some are even able to detect electrical and magnetic fields.

10 The language of our eyes

Of hidden meanings behind our look

9 A world of color

On the meaning of colors

In all animals capable of color vision, color signals play important roles. Their eyes are adapted to prefer specific hues when selecting mates or searching for food. Depending on their number of receptors, they may even experience colors hidden from us.

To some extent, we understand the eye language of our closer animal cousins. This allows us to communicate many simple emotions across different species. The great apes, being our closest cousins on the tree of life, are able to recognize more of these than other animals.

1 Thoughts about evolution

Biological and mathematical ways of seeing

Biological evolution is the change in time through selection, which is the difference in reproductive success of a population due to varying fitness. This mechanism may noticeably alter the frequency of a trait over even a single generation.

Vertebrates …

An evolutionary photo shoot

Evolution still takes place today, throughout the kingdom of all living things, just as it always has since life began on Earth. The results of evolution are visible in two forms: In living organisms, and in what remains of them once they expire – usually, no more than fossils. Eyes comprise such fragile parts of our animal bodies, that they are seldom well-preserved enough to yield many details in a photo (p. 121).

Most nature photographers prefer live motifs, however deceptive their appearance may actually be. From mere pictures, it would be impossible to conclude that whales share recent ancestors with hippos. On the other hand, we often wonder about the apparent lack of important differences among the eyes of vertebrates. The sizes of vertebrates vary profoundly – from tiny shrew mice to great whales, most orders of magnitude in the animal kingdom have their place among them. However, their lens eyes never grow larger than a tennis ball. This is good news for the nature photographer, as this whole range can be covered by a single macroscopic lens!

The inescapability of heritage

Take a look at the evolutionary tree of animals, and compare typical specimens at the branching points. If the theory holds, then there may be no inconsistencies. Eyes have evolved many times, repeatedly and independently – this becomes especially clear as one follows the tree of evolution.

Usually, the emergence of a type of eye implies that the whole evolutionary branch is stuck with this design. However, once a design is evolved, it can be improved or simplified, whichever way the selective advantage lies. A change from facet to lens eyes has only occurred in arachnids and in insect larvae.

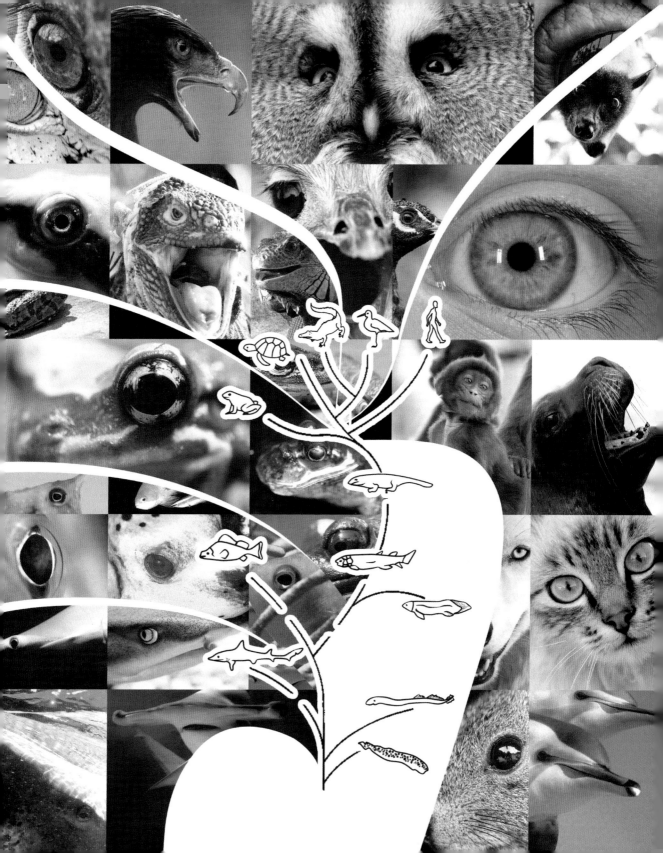

... and invertebrates

The complex system of life forms on Earth, and their gradual development through time, is often depicted in the form of genealogical trees. This is appropriate not only for vertebrates, which tend to be more familiar to the layman, but also for invertebrates, encompassing millions of diverse species: from insects to arachnids and crabs, most invertebrate species have prominent eyes that are used to astonishing effect. For this reason, invertebrates also get their fair share of pages in this book. In order to capture minute details on the eyes of these mostly tiny invertebrate animals, we had to employ modern magnifying lenses. The smallest animals depicted in this book are no more than 2 mm in size, and their eyes are commensurately smaller still (p. 112-113).

Most arthropods have facet eyes, from which spiders, in a secondary feat of evolution, have managed to evolve lens eyes.

Molluscs, encompassing populations as different as octopuses and snails, have no less fascinating lens eyes, of which the eyes of octopi are even comparable to ours! There are whole classes of animals that possess simple capacities for vision to one's surprise. The eyes of sponges and coelentrates (polyps and jellyfish) have, with a few exceptions, little to offer in terms of capability. The pit eyes of flatworms and the tiny eyes of annelids are likewise very simple, and roundworms are blind altogether. Sea urchins, sea stars, and sea squirts have no sense of sight worth elaborating upon. However, they are remarkably close to vertebrates on the tree of life – having contributed gills to their genetic makeup – an ancient feature that humans still recapitulate today during embryonal development (p. 14).

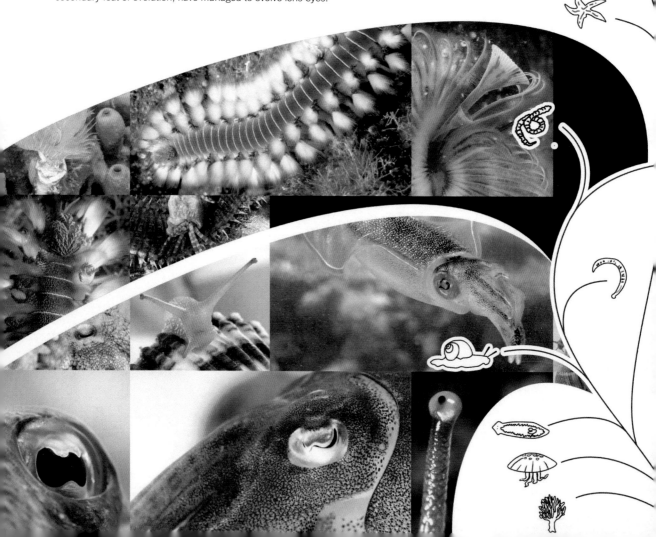

The theory of evolution – A mathematician's point of view

Of definitive proofs
Mathematics is considered to be the strictest of all scientific disciplines – and for good reason. A proof is either complete and flawless, or it is simply not accepted. Throughout history, even the most brilliant mathematicians, like *Carl Friedrich Gauß*, worked on proofs that they were not able to bring to a completion that would pass muster according to today's exacting standards. In many cases, coming generations of mathematicians were able to complete the work of their masters – nonetheless, incomplete proofs still exist with gaps waiting to be filled in.

A single counterexample is enough
Usually, all that is required to bring a mathematical theory to its fall is a single well-reasoned and valid counterexample. For instance, let us consider the famous four color theorem: using only four different colors, it is possible to draw the most complicated map of fictional countries, avoiding that a country painted with a specific color ever neighbors another country painted with the same color (see the image below and the photo on the next page). This claim used to be proven by verifying all thousands of possible color combinations in a computer. When it was found that no counterexample was possible, the theorem was judged to be correct. These days, a more beautiful mathematical proof exists, for which it is not necessary to know all combinations.

Is it possible to prove laws of nature?
In physics, a law of nature is considered "proven" if it the predicted phenomenon is ascertainable under any given conditions, and if no contradictory results are observed. A stone being dropped from a height in complete vacuum is guaranteed to fall towards the center of the Earth, as it has never been observed to react in a different manner under such laboratory conditions. To claim that the law of gravity is not yet proven, after having witnessed a magician's levitation trick, is to invite ridicule from every enlightened person. Yet, a claim of comparable absurdity is constantly repeated with respect to the theory of evolution.

Carl Friedrich Gauß* and *Charles Darwin
Gauß (1777-1855, left image) in his lifetime published only parts of what is now his enormous legacy in mathematics. It was only in 1898, when his diary was discovered, that posterity came face to face with the enormity of his achievements. As early as in his dissertation, he proved the fundamental theorem of algebra which plays a central role in mathematics.
Charles Darwin (1809-1882, right image), a contemporary of Gauß and the founder of evolutionary theory, came to his most significant insights in the 30s of the 19th century. His ideas are fundamental to modern biology, the same way that the ideas of Gauß are the backbone of modern mathematics.

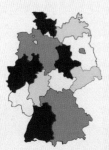   

The squaring of the circle
It has already been proven, once and conclusively, that one cannot construct a circle and a square of precisely equal areas with only a compass and a straightedge (see Leonardo da Vinci's famous drawing). Despite this fact, of which we are mathematically certain, one stills hears of the occasional, unrecognized Internet genius claiming to have accomplished this impossible feat.

Both men were at odds with the opinions of the day. Darwin waited for many years before publishing *The Origin of Species* in 1871, and despite facing constant opposition from theologians and philosophers alike, his theory only grew stronger during the decades hence. Today, after only insignificant changes to its original premise, it represents the "state of science" in all disciplines related to biology.

Four different hues would be enough to color this gorgeous Indonesian cricket completely, without two equal colors touching each other. In fact, this cricket seems to show five colors. Such numbers alone do not tend to hold biological significance, to the disappointment of many would-be numerologists. An overlapping of colors would simply produce different patterns.

Wikipedia http://en.wikipedia.org/wiki/Four_color_theorem
Four color theorem

A further note on the squaring of the circle
A circle cannot be converted to a square of equal area by means of an elementary construction. While this statement has been proven mathematically, a non-mathematician may wonder why it is even useful to attempt such a conversion. Why shouldn't it be possible, and in any case, why does it matter? Such scepticism does not stop people from claiming, occasionally, to have accomplished the impossible feat. It goes without saying that such individuals have little knowledge of mathematics. Usually, they test their invention a few times, and if it comes close to the expected result within a margin of error, they mistakenly believe that this constitutes a mathematical "proof".

In the age of computational geometry, it is relatively simple to refute such untenable claims: one only needs to write a computer program that follows the proposed construction technique to the letter. It then becomes easy to show that the resulting numerical values are not identical to the expected values. As a method of argument, this works better than attempting a mathematical counterproof. Often, mathematical arguments are readily available. Crucially, however, understanding such arguments requires knowledge of higher mathematics, and a lack thereof is what caused the problem in the first place. In mathematics, counterproofs are only valid if they are completely free of error.

Counterarguments to the theory of evolution
Evolutionary theory is able to explain countless apparently puzzling features of the biosphere quite conclusively – much more so than any competing theory. It may be impossible to prove the theory of evolution mathematically, but it is still remarkable that, after more than 150 years of passionate attempts, nobody has yet managed to formulate a cogent and scientifically valid counterargument[1]. Here, we can draw certain analogies to problems of mathematics: Neither the famous Riemann hypothesis, nor the Collatz conjecture, could so far be proven with complete mathematical rigor. However, nobody has yet managed to find a single counterproof. The bottom picture illustrates the Collatz conjecture with respect to the so-called "termination of recursive algorithms". In a certain way, this picture fits perfectly in this book, as it bears a strong resemblance to a genealogical tree.

Let us "prove" that 1 equals 2
and conclude that mathematics is nonsense
Consider another simple example that is often quoted in different forms: The conclusion

$$a - b = 2(a - b) \quad \Rightarrow \quad 1 = 2$$

is, of course, false – otherwise, it would turn the entire edifice of mathematics on its head. The correct statement is

$$a - b = 2(a - b) \quad \Rightarrow \quad a - b = 0$$

since all multiples of zero are still zero. The first conclusion's mistake lies in the cancelling operation. For this precise reason, divisions by zero are not allowed in mathematics.

Nobody would seriously object to the above statement – after all, mathematics forms the basis of our technical accomplishments, and without it, natural science would be impossible. A reasonable person would not be tempted to set up a blog where the whole of mathematics is rejected based on such an obviously erroneous and easily refutable argument. However, the Internet is rife with equally laughable "counterarguments" attempting to disprove the theory of evolution. To give a mathematical analogy, these individuals appear convinced that multiplying zero by a large enough number will eventually lead to a non-zero result.

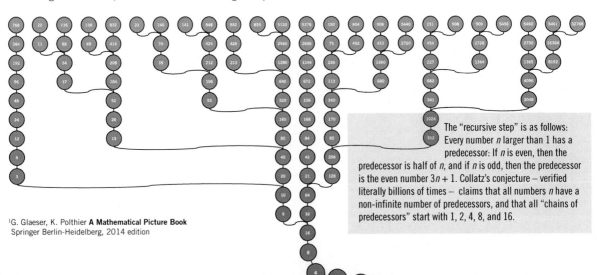

The "recursive step" is as follows: Every number n larger than 1 has a predecessor: If n is even, then the predecessor is half of n, and if n is odd, then the predecessor is the even number $3n + 1$. Collatz's conjecture – verified literally billions of times – claims that all numbers n have a non-infinite number of predecessors, and that all "chains of predecessors" start with 1, 2, 4, 8, and 16.

[1] G. Glaeser, K. Polthier **A Mathematical Picture Book** Springer Berlin-Heidelberg, 2014 edition

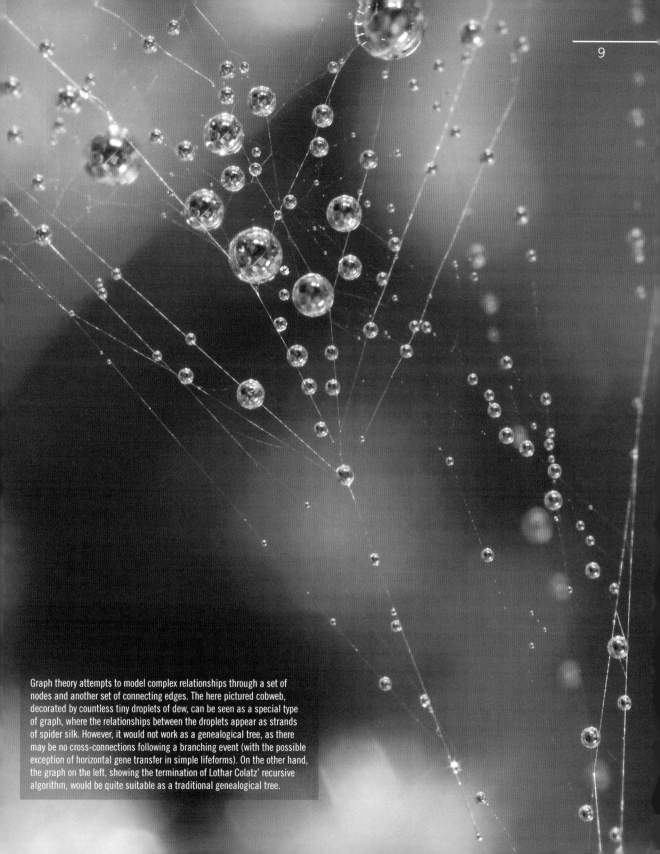

Graph theory attempts to model complex relationships through a set of nodes and another set of connecting edges. The here pictured cobweb, decorated by countless tiny droplets of dew, can be seen as a special type of graph, where the relationships between the droplets appear as strands of spider silk. However, it would not work as a genealogical tree, as there may be no cross-connections following a branching event (with the possible exception of horizontal gene transfer in simple lifeforms). On the other hand, the graph on the left, showing the termination of Lothar Colatz' recursive algorithm, would be quite suitable as a traditional genealogical tree.

A biologist's point of view

Palaeontology as a window into the past
At least since *Jean-Baptiste Lamarck* (1744-1822), it was known that organisms undergo change over the course of evolutionary time. We know from paleontology that older layers of rock contain fossilized animals of a more primeval nature, with vertebrates appearing only in later strata, following the *Cambrian explosion*. The remains of mammals and birds make their first appearances in even younger layers. The strata in the Earth's soil, which can be dated with remarkable accuracy, offer a glimpse into the distant past of our planet's history. In his ground-breaking *Philosophie Zoologique* from 1809, Lamarck was only able to show the existence of evolutionary processes – however, he was still unable to explain what caused them.

Evolution is the change of organisms in time due to selection
50 years after Lamarck's discovery, *Charles Darwin* managed to explain evolutionary changes and their causal factors:
1. *Biological evolution* means that all populations of organisms experience a change in their phenotypes over time.
2. These evolutionary changes occur exclusively in small steps, representing the difference between the parent organisms and their offspring.
3. The diversity of species stems from the branching of genealogical lines. It occurs in addition to evolutionary changes within these lines.
4. The mechanism that controls the diversification of mutations and that imprints itself on genealogical history most dramatically is *natural selection*, which Darwin described in analogy to *artificial selection* – the selective breeding of animals and plants by humans.
5. All organisms share a single common ancestor. The manifold diversity of organisms is the product of a genealogical development measuring billions of years, following the emergence of *chemical evolution*. Since this development must have started with a single species, all organisms on Earth are related.

Establishing a theory of selection
The ideas of Darwin still represent the only scientifically verifiable theory about the origin of life, and stand in stark contrast to the myriad of creation myths that are believed by different cultures throughout the world (*Genesis* being just one of many). Most of these mythologies assume a singular creation event, and a subsequent constancy of species in the biosphere.

Darwin's theory has famously sparked endless debates that still continue today – it is remarkable, however, that virtually none of these debates are based on the scientific method, including the explanations from the "intelligent design" school, which are largely pseudoscientific.

Artificial selection: waiting for specific mutations
Darwin assumed that the natural selection within populations of organisms may, on average, lead to an improvement of their biological or economical attributes. The technique of selecting and breeding animals and plants that – by pure chance – possess desirable traits was widely practiced in Darwin's day and is still being practiced today. Through continued selection over multiple breeding cycles it is possible to arrive at a breeding goal, whatever the goal may be. Breeders of ages past used to wait for so-called "hot spots" and continued to breed further offspring from such individuals. Today, we know that these "hot spots" are actually mutations – changes in hereditary information within germ cells.

**Natural selection:
the difference in reproductive success of individuals**
Darwin correctly recognized that selection also occurs in nature and that, in principle, it is no different from artificial selection. Natural selection is the difference in reproductive success of individuals in a population. This difference follows from varying genetic "fitness" – however, while the differences among individuals in a population are due to random mutation, the process of selection itself is not random at all, but results from the genetic constitution of each individual organism. It makes its mark on all natural populations, influencing the genetic repertoire of coming generations. Individuals more successful at reproducing – for whatever genetic reason – increase the frequency of similar genetic traits in subsequent generations.

Survival of the fittest:
successful mutants displace the less successful ones
Populations in nature tend to have a roughly constant size. For this reason, genetic traits that cause their hosts to be less successful at reproduction tend to disappear over time. Darwin called this the *"survival of the fittest"* – a phrase that is often misunderstood. It does not always refer to tooth-and-nail fighting, or to a naive conception of physical strength, but to a contest that is won or lost through higher or lower reproductive success. A marathon race is a good analogy. It isn't the strongest or the most violent individual that wins, but rather the one that runs fastest, by average, across of 26 miles. In evolutionary

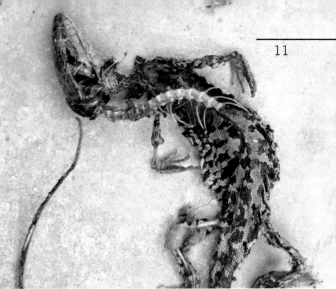

comprise the myriads of interactions with other organisms – most dramatically, in the competition for natural resources.

Two-gender reproduction and meiosis
Selection is only purposeful insofar as the individuals of a population differ from each other in certain genetic traits. Among nature's earliest tricks to jumpstart selection was the emergence of sexuality with two genders, combined with a means of cell division known as meiosis. It differs from traditional cell division (mitosis) in that the chromosomes are first split in half before being recombined, leading to a unification of the parent chromosomes.

A virtually infinite number of germ cells
Mitosis, and its combination of hereditary information from two parent organisms, produces a germ cell – a so-called gamete. This has interesting consequences. Assuming roughly 1000 structural genes and two forms that a gene (allele) may take within the locus of a chromosome, there exist more than 2^{1000} possible combinations – an enormous number with more than 300 digits!

biology, this would mean that only the winner or the group of winners would be able to reproduce, and that this privilege would be upheld across many generations.

Biotic and abiotic factors of selection
Natural selection is strongly influenced by biotic and abiotic factors of the environment in which a population lives. Abiotic factors include temperature and humidity, while biotic factors

Every egg and sperm is unique

No matter how many hundreds of millions of sperm a male is able to produce during ejaculation, no pair of sperm are genetically identical. Even if all the billions of men on Earth are counted, they will never be able to produce two genetically identical pieces of sperm or spermatozoa. The same is true for the female egg. This demonstrates that recombination is a motor for practically infinite variety.

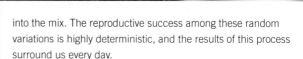

into the mix. The reproductive success among these random variations is highly deterministic, and the results of this process surround us every day.

The evolution of optical sensory organs

If the possession of better optical sensory organs grants an organism a relative survival advantage, then a statistically higher number of descendants will inherit this successful attribute. If the selective pressure is very large – such as in questions of detecting a predator with speed and accuracy – then good optical sensory organs may spread relatively quickly among the population, and will also keep improving over time.

Channeling boundary conditions

The type of eyes that evolve depends on many preconditions, such as the environment in which the population lives. The physical attributes of the environment constantly enforce

Selection may be a static process, but it is the opposite of randomness

In addition to the dazzling number of chromosomic recombinations, mutations increase the diversity of variations even further. This produces a very practical scenario for selection, as it is able to consider a virtually infinite amount of slightly varying organisms. However, selection is a static process – just as in the throwing of a dice, the singular event is of little importance, and patterns only start to emerge after many attempts. The only truly random element in the selective process is the diversity of available variations – that is to say, which structural genes are combined during meiosis, and which mutations are added

channeling boundary conditions – for instance, the production of a sharp image on a retina is dependent on the optical laws of refraction. For such reasons, eyes that are predominantly used at night are constructed differently from those that are usually applied during the day.

The successive evolution of different types of eyes

Our distant ancestors had to evolve successive techniques and constructions for the purpose of light perception, and for the subsequent signal processing through the nervous system. In fact, through the study of contemporary animals, it is possible to arrange an order in which our eyes must have evolved – from the humblest of beginnings to the most sophisticated lens eyes.

Detecting light and darkness

The earliest form of the eye represents no more than a single cell with a photo-sensitive protein pigment. Photo receptors absorb the energy of the incoming photons with a receptor protein, which changes the conformation of carbon atoms – the spatial arrangement of the rotating attachments. A form of signal processing thus emerges, wherein ion channels within the membranes are opened and closed, leading to a change of the membranic potential within the protoreceptor cell. Single cells of this type are initially only able to discern light from darkness, or very simple shadow phenomena.

A combination of modules

The detection of directional light is accomplished by adding another cell with a light-absorbing pigment. This very basic equipment – the precursor to all later types of eyes – appears to have evolved very early in the history of life on Earth. From this basic module, different types of eyes have subsequently evolved, and often completely independently of each other.

The embryo and all stages of the eye's evolution

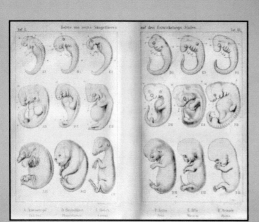
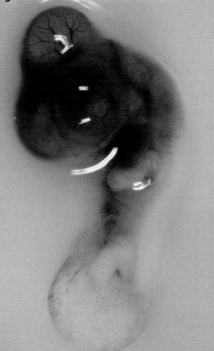

The zoologist *Ernst Haeckel* (1843-1919) was among the earliest and most passionate advocates for the theory of evolution in the German-speaking world – in fact, he extended Darwin's idea to form a special hereditary theory of animals. As a comparative anatomist, he was inspired by the observations of the Estonian naturalist, anthropologist, and embryologist *Karl Ernst von Baer*, who by that time had already noticed that the embryos of mammals are remarkably similar to each other. By extrapolating Baer's discovery, *Haeckel* formulated his "biogenetic law" that is cited to this day, stating that a recapitulation of phylogeny occurs during ontogenesis. He postulated an evolutionary causal relationship between embryological development and the phylogeny of a species, and visualized this relationship through a sketch that is still famous today, showing the embryonic stages of eight different vertebrates – including fish, amphibians, reptiles, birds, and humans. *Haeckel* stated that these embryos share a strong resemblance during early developmental stages, with the most dramatic differences occurring only later. From this observation, he concluded that the embryos of the first vertebrates must have looked like the contemporary common form at the start of embryonic development. While this statement later turned out to be too simplified, the basic assumption is correct: the early traits of embryos are also subject to selection. Today, these ideas have produced their own branch of biological science – developmental evolutionary biology, often abbreviated as "EvoDevo".

Today's understanding of biogenetic law builds upon the observation that an organism is a system that is subject to perpetual change and reconstruction.

E. Haeckel **Generelle Morphologie der Organismen** 1866

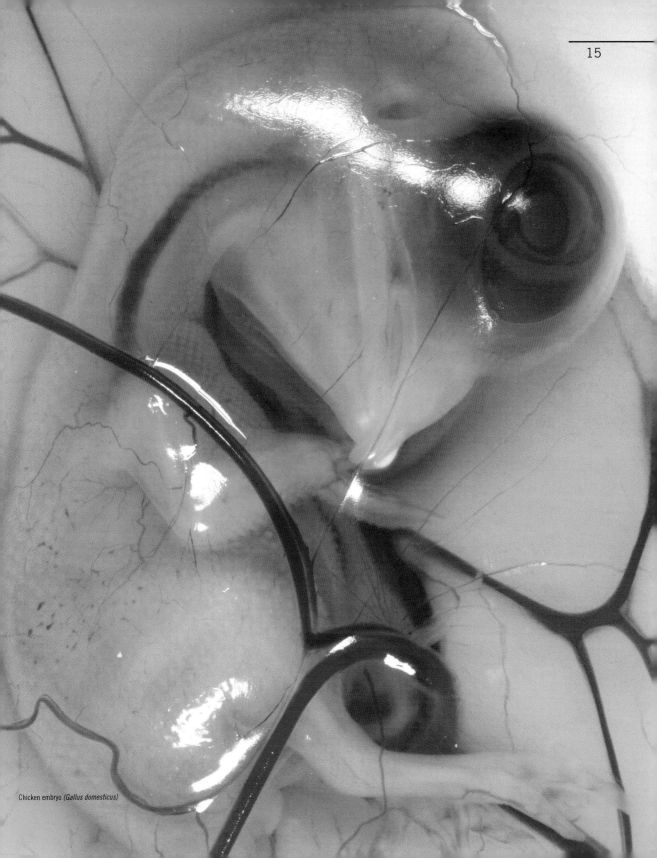
Chicken embryo (*Gallus domesticus*)

2 Lens eyes or compound eyes?

Two very different ways to see the world

The light-sensitive cells of all eyes — no matter their difference from ours — evolved either from the nervous system, or from cells of the skin. It only took those two types of cells to produce lens eyes and facet eyes, although only after a large number of generations.

Lens eyes or compound eyes?

The 500 million year old legacy of all animals
Both lens and facet eyes were "invented" multiple times throughout animal history. Lens eyes first emerged in the Paleozoic in vertebrates and octopi. However, lens eyes had already existed within cnidarians – mainly within highly poisonous box jellyfish. As fossils, these animals make their earliest appearance in the mid-Cambrian, or about 500 million years ago, but because other medusa species among the cnidarians also possess small lens eyes, it stands to reason that these must have emerged prior to the Cambrian. In other words, eyes, and lens eyes in particular, represent an ancient legacy of all animals.

Facet eyes, on the other hand, form the base plan of all arthropods and are today widely distributed among crabs and insects. Their marine ancestors must have had eyes for at least 555-600 million years, as they already feature so prominently in the long extinct Trilobites. Similar types of eyes can sometimes be found in bristle worms (Polychaeta). According to this theory, the wide diversity of eyes evolved at least 500 million years ago – albeit within a relatively short span of time.

Convergent evolution of lens eyes
Eyes or structures with a similar function appear in almost all groups of animals – ranging from simple, light-sensitive *collections of cells* on the epidermis (in earthworms), cup-shaped eyes without lenses (in Turbellaria), pinhole camera eyes (in the Nautilus), to eyes with a refractive lens that project an image of the environment onto a curved retina. Throughout evolutionary history, each of these types emerged multiple times and independently of each other, nevertheless producing remarkably similar results. This phenomenon is known as convergent evolution and can be observed in earthworms, snails, octopuses, arachnids, and especially in vertebrates.

Small eyes or large eyes?
According to the laws of optics (which also apply to photography), lenses with a smaller focal length have a greater depth of field, but also less space for the imaging sensor. High-resolution eyes with a large focal length require refined methods of focussing and accommodation – especially if they are also large.

Focussing with a corrective lens
If the visual image is to be sharp, larger animals possessing lens eyes with large focal lengths must, therefore, repeatedly refocus their eyes as their gaze moves across objects at varying distances. Cartilaginous fish, ambhibians, and snakes move a static lens forwards in order to capture close proximities, as the "default position" of their eyes is intended for looking at distant objects.

Bony fish have the opposite approach. They move a static lens backwards in order to focus on distant objects, as their "default position" is suited for objects close by. The lenses of higher animals undergo deformation, changing their refractive index and focal length. Birds and reptiles accomplish this through direct pressure from the ciliary muscle. Mammals deform their lens more indirectly through so-called zonule fibers within the aperture of the ring-shaped ciliary muscle. Pulling on these fibres causes a flattening of the lens. If the fibres are to be relaxed

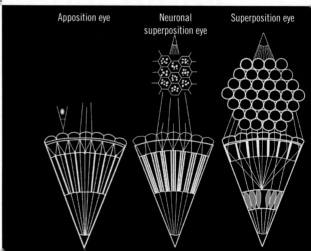
Apposition eye   Neuronal superposition eye   Superposition eye

for close-up focus, the ring-shaped muscle must contract, as the lens possesses an elasticity of its own which tends to keep it in a spherical shape.

Reflector eyes – an interesting variation upon lens eyes
Reflector eyes, of which scallops of the genus *Pecten* possess 60-100 instances, function according to a different imaging principle. Similar mirror optics were recently discovered within barrel eyes, an example of which is found in Opisthoproctidae, a species of deep-sea bony fish. The focal point of a reflector eye lens lies far behind the plane of the retina – thus, the light rays are projected onto a double-surface retina from behind through a concave mirror, which acts as a reflecting surface.

Complex and facet eyes of crabs and insects
In complex eyes, a large number of ommatidia project the environment onto a complex retina, with every individual unit possessing a small system of lenses. Each ommatidium, usually made up of eight elongated sensory cells, forms a part of this retina – a so-called *retinula*. In most cases, the retinula is optically separated from its neighbors through so-called screening pigment cells. Apart from that, each visual unit contains its own dioptric apparatus in the form of a transparent chitin-based cornea, supported by an underlying four-part crystalline cone. Both have only a very short focal length. This makes facet eyes suitable for extreme wide-angle imaging – just as the fixed focus lens of a simple pocket camera is able to focus on practically anything beyond a few millimeters, the facet eye accomplishes a similar feat, but with many hundreds or even thousands of such miniature cameras arranged in a semicircle.

The structure of an ommatidium
At the entrance of each individual facet, a chitin-based cornea focusses incidental light parallel to the longitudinal axis of the eye. Passing through the bottom opening of the crystalline cone below, such light impacts upon the surface of eight optical cells. This means that each system of lenses, accompanied by its retinula, forms a self-contained visual unit – an ommatidium. The light pigments are contained within the membranes of each retinular cell. In order to enlarge the membrane surface, finger-shaped extrusions appear across almost the whole length of the optical cell – so-called *rhabdoms*. Rhabdomeres and rhabdoms together form an optical fibre that captures and transmits the incident light towards the outbound axons of each optical cell. Scattered light is absorbed by a hull of tiny pigment cells that surround each ommatidium. Such a facet eye that is based on optically isolated ommatidia is called an *apposition eye*.

Resolution and luminosity
The diameter of a lens is essential, as it determines the size of the area though which photons may incide. Large lenses are very light-sensitive for this reason – however, due to their short focal lengths, they are only capable of limited resolution. A focusing of light through small lenses or through an aperture (in the case of photo cameras) increases visual sharpness at the expense of luminosity.

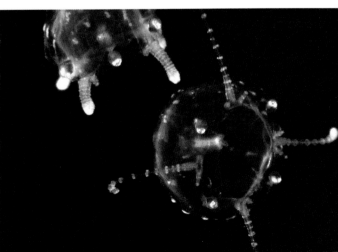

Facet eyes for day and night

Apposition eyes with their optically isolated ommatidia provide relatively good sight to diurnal insects and crabs. Nocturnal insects have evolved so-called *superposition eyes*, in which the pigment cells surrounding each ommatidium are reduced in the direction of the retinula. Apart from that, the focal length of such lenses is slightly larger, leading to a focussing of light through adjacent ommatidia onto the shortened Rhabdomes. This makes such eyes more sensitive to light, while also reducing the sharpness of the image. Eyes of this type can be found in moths (see left photo).

The large pseudo-pupil (p. 180) on the eye of the moth suggests a superposition eye.

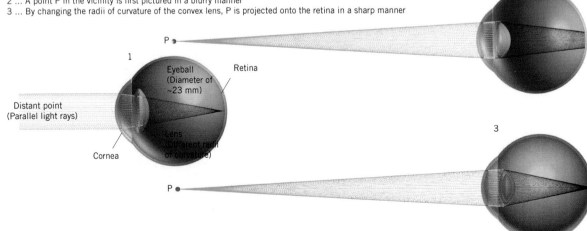

Simulation of refraction in the human eye (according to *Allvar Gullstrand*):
1 ... The system focusses parallel incident light onto the retina
2 ... A point P in the vicinity is first pictured in a blurry manner
3 ... By changing the radii of curvature of the convex lens, P is projected onto the retina in a sharp manner

Each individual eye only perceives a fragment of the environment, but when the visual information of all facets is put together, a complete – if slightly rasterized – picture emerges.

An overstated difference?
Despite the different techniques at play in lens and facet eyes, certain principles are shared in common. The axons of the optical cells encode the visual information as electrical signals and forward it to the nervous system. A neuronal image only appears after these signals are processed by the brain. Thus, it would not be completely mistaken to assume that despite the difference between the mechanisms that capture visual information, the cognitive representation of the animals' environments and of the objects within them might be more similar than one would intuitively assume, judging merely by the construction of the eyes.

The eyesight of articulate animals

The number of facets (ommatidia) plays a similar role in determining the sharpness of the image as the number of pixels in a photograph. Each individual facet captures a single point of the image. A honeybee possesses 5.000 ommatidia on each side of its head whereas a housefly has only 4.000. Some larger dragonflies can have up to 30.000, while some ants' vision is limited to 6 ommatidia. Lenses and optical cells are located closely behind one another on a mostly convex (spherical) surface. Despite a weakness of resolution, flying insects are capable of maneuvering remarkably precisely at great velocity. This is possible through the principle of *hyperacuity* whereby the overlapping fields of vision of adjacent facets yield additional information that leads to a more accurate representation of the environment in the brain.

Tiger fly *(Coenosia tigrina, Muscidae)*

Good eyes are a requirement for hunting

Both images on this spread depict the same prey – a dance fly (Empididae), an insect as small as two millimeters. The predators, however, are quite different. The left side shows a tiger fly (*Coenosia tigrina*, Muscidae), a distant relative of the common housefly. A zebra spider *(Salticus scenicus)* appears on the right. It is not uncommon for flies to prey on small airborne insects, as their agility during flight equips them well for this task. Their eyes, which perform very well at close range, are equally beneficial for hunting. As with all jumping spiders, the anterior eyes are especially large and directed forwards in the same direction as the smaller pair. The remaining four eyes enable the spider to see backwards. Zebra spiders are capable of focusing on tiny distances impossible to see for the human eye. As opposed to other spiders, which rely on the detection of vibrations and movements, jumping spiders are also able to detect motionless prey (such as dead animals).

Zebra spider *(Salticus scenicus,* Salticidae*)*

Image raising through refraction

The different indices of refraction in the separate layers of the cornea produce remarkable refractive phenomena – a challenge for the photographer, but also for the scientist keen on drawing conclusions from a photograph.

The series of images on the right shows a human eye, whose iris is shaped like a planar ring. When looking at the eye from the front, the coloring of the iris appears to have an effect on the entire anterior hemisphere. If the optical axes of the camera and the eye are at right angles to each other, then the vitreous body becomes transparent, and the outer cornea comes into view. Similar effects are visible in the *computer rendering*. This book (including its cover) quite deliberately contains many depictions from perspectives that are photographically interesting, and aesthetically pleasing. The drops of water on the eyes of the fly below show the optical phenomenon of "image raising" that also produces a remarkable and dramatic magnification effect.

Housefly *(Musca domestica)*

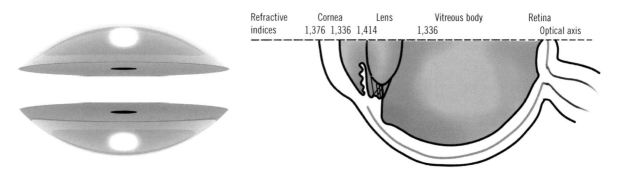
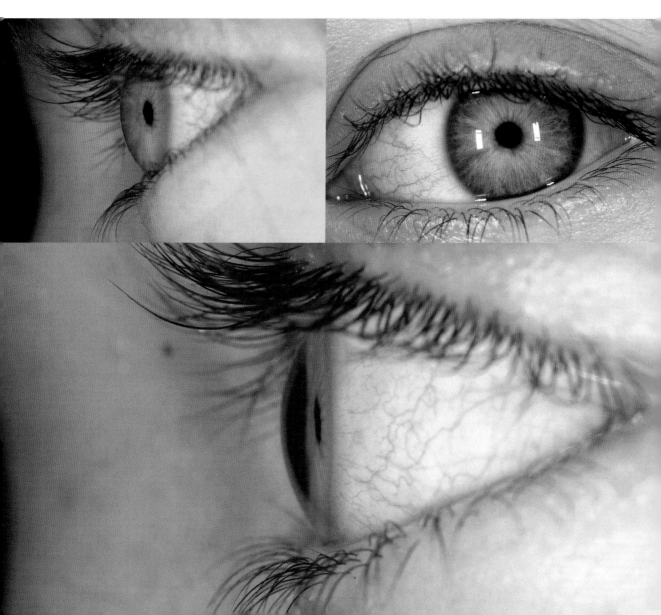

Two solutions for sharp eyesight

Ecuadorean jumping spider (*Salticidae*)

In the right photo, an assassin fly from the family of Asilidae has captured an insect and proceeds to extract liquids from its body. Aided by large spheres covered in thousands of ommatidia, it was able to get a sharp image of its prey. In the left photo, a jumping spider from the family of *Salticidae* has taken hold of a prey of its own – a smaller spider. Spiders do not, in fact, have facet eyes, but 8 separate lens eyes – the anterior ones (called the "main eyes") being especially large relative to the animal's body.

Robber fly *(Dioctria linearis)*

Advantages and disadvantages of the facet eye

Excellent spatial vision!

Yellow-legged Robberfly *(Dioctria linearis)*

Quick reaction times of facet eyes

In the world of tiny distances, facet eyes have several clear advantages when compared to lens eyes. For us, a sequence of images at 20 frames per second is enough to give an impression of fluid motion. Insects, however, are capable of seeing 300 separate images per second. This gives them extremely quick reaction times. Apposition eyes are capable of a temporal resolution of about 80 frames per second. Furthermore, animals with facet eyes have a much wider field of vision than those with lens eyes. In principle, each separate facet is capable of the same close-up sharpness over a wide field of view. As we know from our own experience, lens eyes only let us focus on a limited area at the center of our field of vision.

Fixed eyes and a flexible head

The striped Yellow-legged Robberfly *(Dioctria linearis)* on the left, measuring only about one centimeter, turns its head towards the camera. From this behavior, we can conclude that it is capable of measuring distances more precisely towards the front of its field of view. The photos below depict a tarsius *(Tarsius tarsier)* from Sulawesi, which is no more than 11 cm in size. The animal does not move its eyes spherically, as would be customary for other primates, but instead rotates its head, which sits on a short but very flexible neck.

Tarsius *(Tarsius tarsier,* Tarsiformes*)*

The spatial vision of insects

Bush cricket *(Tettigonia viridissima)*

Japanese Robberfly
(Dioctria nakanense)

Grasshoppers (left page) jump and fly across short distances. It thus stands to reason that they, like all airborne insects, possess good vision. As herbivores, their eyes are positioned at the side of their head. This maximizes their field of vision and permits them to detect predators more quickly. On the other hand, the assassin fly above has the essential parts of its eyes pointed forwards. Notice the bicolor appearance of its facets, which aids the insect in camouflage (somatolysis).

From a mathematical point of view, the distribution of hexagons on a spherical insect eye is no trivial problem. It can be shown that no exact solution exists, which leads us to the conclusion that the hexagons must be either irregular, or slightly different in size. The illustration on the right shows the central view axes of select facets. One immediately notices that certain pairs of axes intersect each other, denoting spatial coordinates for a fixed number of points. Remarkably, this pairwise relationship aids the animals in spatial vision. Close-ups of insect eyes rarely go close enough to allow us to discern individual hexagons – however, it is still possible to notice that most conical facets are surrounded by six neighbouring ones (see photo above). Our perception automatically interprets this relationship in terms of a hexagon, just as we are prone to see spirals in the arrangement of seeds on a sunflower[1]. With the aid of an electron microscope, we are actually able to see hexagonal edges separating the facets.

[1] G. Glaeser **Nature and Numbers**
Birhäuser / De Gruyter, 2013

Crab or insect?

The European mole cricket *(Gryllotalpa gryllotalpa)*, indigenous to all of Europe and Northern Africa, represents a family of long-horned grasshoppers within the class of insects *(Insecta)*. They are close relatives of the more frequently encountered field crickets.

The name "mole cricket" is due to these animals' characteristic appearance. On the one hand, they wield fearsome shovel-like claws, burrowing into the soil like moles. On the other hand, they resemble large crickets in their general shape, and even produce similar sounds.

Judging by their appearance, one might mistake them for crabs. However, there are important differences between insects and crabs. Insects have only six legs, while higher crabs have ten. What is more, crabs have movable stalk eyes, while the facet eyes of insects are in a fixed position on their heads.

European mole cricket *(Gryllotalpa gryllotalpa)*

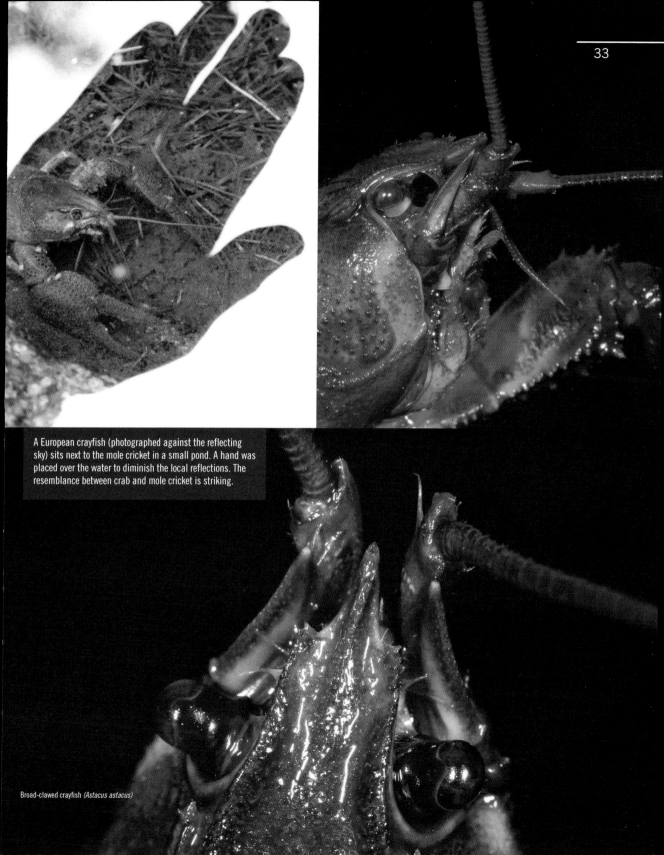

A European crayfish (photographed against the reflecting sky) sits next to the mole cricket in a small pond. A hand was placed over the water to diminish the local reflections. The resemblance between crab and mole cricket is striking.

Broad-clawed crayfish *(Astacus astacus)*

Rolling eyeballs

Wikipedia **Strabismus** http://en.wikipedia.org/wiki/Strabismus

Marmalade hoverfly *(Episyrphus balteatus)*

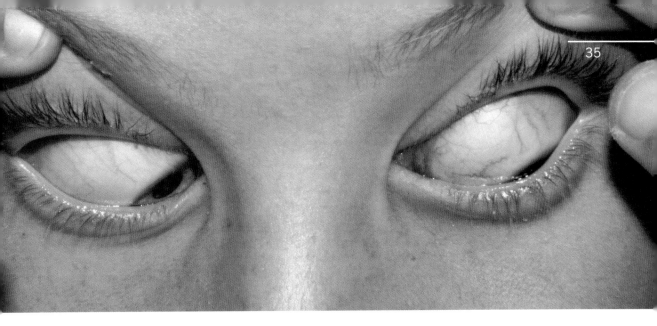

To roll one's eyes …

… is more than just child's play, and in fact, quite important to our survival. Chameleons and some species of fish are also capable of rotating their eyes to face different directions. Crabs can move each of their stalk eyes independently. While it is still unclear whether crabs actually use both stalk eyes as they point in different directions, the situation is simpler with insects, whose facet eyes are completely fixed. However, insects accomplish an analogous effect by turning their entire heads. The hoverfly on the left page manages this feat by a remarkable 180°.

Cross-eyed seeing – a defect of eye muscle equilibrium

Unless one is squinting on purpose, cross-eyed vision is simply the temporary or permanent deviation of vision lines from the object being observed. The degree and form of this malposition may vary, but the so-called "squint angle" can still be measured with relative accuracy. Some forms of cross-eyed seeing are simply a physiological deviation from the ideal. Other forms do not just cause a cosmetic problem, but represent a serious affliction leading to a significantly reduced capacity for eyesight. Not least in non-verbal communication, our eyes and their mutual positioning transmit important social signals. For this reason, cross-eyed individuals sometimes make us uncomfortable, as we are less able to intuit their direction of sight.

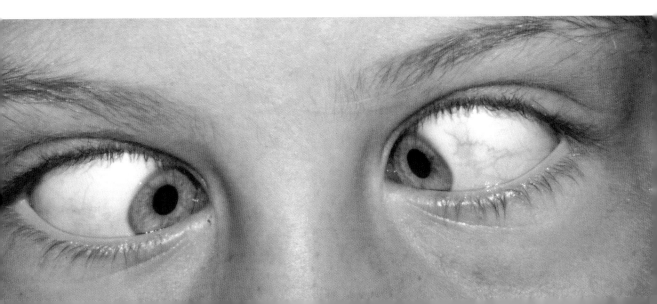

Spider or crab?

Horseshoe crabs are marine representatives of the family *Limulidae*. Despite their name, they count among the most primordial of arachnids, which are set apart from other arthropods through a special pair of feeding claws – the so-called chelicerae. The whole taxonomical group – the Chelicerata – is named for these instruments.

From the ancient precursors to these marine animals, a group representing the ancestors of today's scorpions went out to colonize dry land, being originally nocturnal. In due course, they have reduced their original facet eyes to such a degree that only sparse elements remain – 3 to 5 lenses on either side of their bodies.

Once the transition to land had been accomplished, other groups or terrestrial arachnids evolved, such as the harvestmen, the mites, and the spiders. In this sense, horseshoe crabs can be seen as living fossils. They are the only representatives of ancient marine arachnids, and they show us that all arachnids descend from ancestors that already had facet eyes.

Facet eye of a *horseshoe carb*. The ommatidia are arranged in a rather irregular manner. They do not have a crystalline cone below their strongly convex cornea, but they somewhat make up for it by an increased number of retinula cells.

The horseshoe crab is a modern representative of primordial marine arachnids. As in all animals of this type, the body is segmented into two larger sections — an anterior (prosoma) and a posterior (opisthosoma). In its ancestral form, the sting served as a precursor to the venomous scorpion's tail.

Real crabs, such as this colorful hermit crab *(Dardanus callidus)*, are equipped with very flexible stalk eyes that counteract the primary disadvantages of fixed facet eyes. This allows crabs to glance in different directions without the need to rotate their head.

Strange eyes

Red rock crab *(Grapsus tenuicrustatus)*

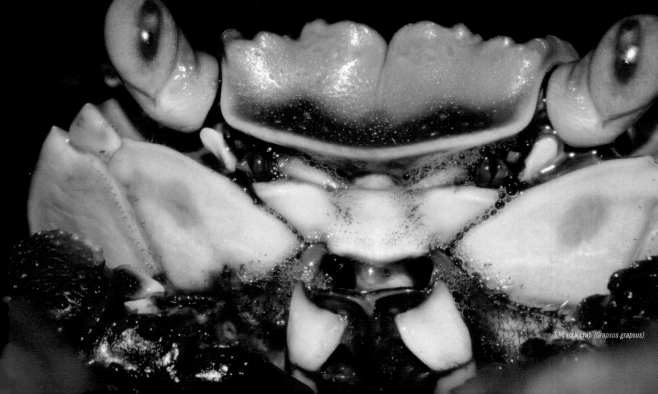

Red rock crab *(Grapsus grapsus)*

Amphibian crabs, such as the common shore crabs of the genus *Pachygrapsus*, have a wide anterior body, which provides the flexible facet-based stalk eyes with 360° view and a wide binocular perspective. The crabs retreat to their self-dug holes to escape from floods or predators.

The eyes of the mantis shrimp (Stomatopoda) are especially bizarre. Its stalk eyes are either spherical or oval, but they always exhibit a band of six ommatidia rows at their center, which subdivide the facet eyes. This band counts among the most sophisticated sensors in the animal kingdom, capable of discerning over 100,000 colors – including ultraviolet and polarized light. In fact, these animals communicate by means of colored light signals. While the upper and lower halves are dedicated to the detection of shapes and motion, the highly complex sensor in the midsection handles the detection of colors and polarization. Its field of view measures roughly 10-15 degrees. Due to the independence of the separate stalk eyes, the crab is able to use one eye to analyze an object's form, while simultaneously using the other eye to inquire about its color and polarization features.

The first four rows are specialized for color perception. Each row contains eight types of visual pigments, ranging from 400 nm (violet) to 550 nm (green). Using adjacent rows, the animal is able to detect linearly polarized light. In part, this is accomplished by orienting the visual pigments in the separate eyes in different directions. This aids in the detection of contrast – an essential feature in underwater environments which are low on contrast to begin with.

Mantis shrimp (*Stomatopoda*)

Facet eyes below the water

The ancestors of contemporary arthropods originated in the ocean – and thus, it follows that facet eyes must first have evolved below the water. Many of these ancestors became extinct towards the end of the Permian – the then-omnipresent trilobites being among the species that disappeared. Among such early forms, only the horseshoe crabs (*Limulidae*) remain in today's biosphere. Their facet eyes are especially simple, as their ommatidia do not possess crystal cones. The ancestors of crabs and insects evolved a four-part crystal cone and a retinula made up from eight sensory cells inside the ommatidium.

While the crabs remained in the water, save for a few exceptions, the insects went on to conquer dry land for themselves. At first, they possessed apposition eyes. For the sake of increased sensitivity to light, some of these eyes must have gradually changed into superposition eyes, at the expense of resolution. Higher crabs have evolved an even more clever way of seeing. In place of crystal cone lenses, they possess rectangular mirror optics, which direct incident light at the multilayer inner surfaces near the top of the rhabdom. These mirrors are positioned exactly perpendicular with respect to each other. The cornea lens is likewise rectangular, as opposed to the hexagonal design of regular facet eyes. With this system of mirror lenses, the animals are able to picture their surroundings with a high sensitivity to light and with a high resolution, even as their eyes stay relatively still.

The manner by which the images emerge is extraordinarily complicated. Through an ingenious system of mutually penetrating virtual mirror surfaces that form conic lateral surfaces around the central ommatidium, a so-called "omnidirectional mirror system" is produced. A concentric pencil of virtual mirror surfaces surround each spatial direction. Due to their virtual nature, these surfaces are able to penetrate the pencils arbitrarily, enabling – simultaneously – a high spatial resolution and a high aperture. This system of lenses was first discovered by *Klaus Vogt* in 1980.

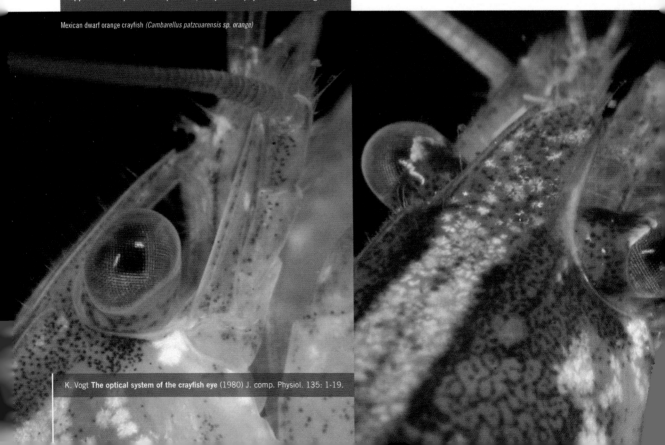

Crayfish (Astacidae) count among the higher crabs with mirror superposition eyes. From the depicted point of view, their corneas and crystal cones appear to be square in shape. Thus, the pseudo-pupil is also rectangular.

Mexican dwarf orange crayfish *(Cambarellus patzcuarensis sp. orange)*

K. Vogt **The optical system of the crayfish eye** (1980) J. comp. Physiol. 135: 1-19.

Marbled shrimp
(*Saron marmoratus*, Hypolitidae)

This colorful specimen from the family of cleaner shrimps inhabiting the tropical indopacific also has eyes on stalks, despite being hard to see due to somatolysis.

Newts and salamanders

Young common newt female *(Triturus vulgaris)*

Newts and salamanders belong to the caudates among amphibians. Usually, they experience their development as larvae in the water. In order to accomplish the transition to land during metamorphosis into the adult stage, a few alterations are necessary. Gills, for instance, are converted into lungs. The visual system likewise undergoes important changes. In the larval stage, tadpoles absorb light on their retina by means of porphyropsin (like freshwater fish), while adult animals use rhopsodin (like land vertebrates). Due to the porphyropsin's propensity for the absorption of long-wavelength light, we might conclude that this represents a special adaptation for underwater vision.

W. Himstedt **Das Elektroretinogramm des Feuersalamanders *(Salamandra Salamandra L.)* vor und nach der Metamorphose** 1970, Pflügers Archiv 318: 176-184
G. Luthardt, G. Roth **The Interaction of the Visual and the Olfactory Systems in Guiding Prey Catching Behaviour in *Salamandra Salamandra L.*** 1983, Behaviour, 83, (1-2): 69-79

The Axolotl (Ambystoma mexicanum) is a Mexican caudate from the family of mole salamanders (Ambystomatidae) that lives exclusively in the water. Remarkably, the animal is only encountered in its larval stage and achieves sexual maturity without leaving this developmental form. This particular species reaches adulthood without metamorphosis, and without undergoing significant changes to its morphology. It retains its gills for all of its life.

The name Axolotl stems from the Aztec language Nahuatl and refers either to a "water monster" or – in a different translation – to a "water doll". In nature, these animals are only found in a few lakes near Mexico City. They are famous for their regenerative abilities, which are widely tested in laboratories of biology. These animals are able to completely grow back extremities, gills, and even their brain and heart!

Olm *(Proteus anguinus)*

Fire salamander *(Salamandra salamandra)*

Fire salamanders are mostly active at twilight, for which they require excellent vision. It was shown that they are able to discern prey at luminosity levels of only a few Lux. In addition, their ability to return to their home location is extraordinary – indeed, individuals may be found for years upon years in the same spot, although this skill might be supported by their sense of smell as much as by their vision.

Reptiles and amphibians in the same biotope

Grass snake *(Natrix natrix)* swallowing a water frog

A grass snake hunts for pray in a frog pond. As a reptile, it requires food only comparatively rarely. Remarkably, it smells its prey with its tongue! The way in which its jaws open as it engulfs its unfortunate victim is reminiscent of the moray (p. 48), though they are not closely related. In evolutionary history, reptiles evolved from amphibians, which themselves evolved from bony fish, to which the moray still belongs.

The eyes of both animals share a strong resemblance. For the purpose of protection, the frog pulls its third eyelid over its eye from below. Alas, this does not ultimately alter its fate.

The eyes of frogs

Indonesian shrub frog *(Polypedates spec., Rhacophoridae)*

Frogs are unable to move their eyes, and if they wish to focus their gaze on a prey animal, they need to reposition their entire body towards it. By default, their noticeably spherical lens is set towards a focus on infinite distances. During accommodation, the lens is pushed forward, and the focal point is moved closer to the animal. Their retina contains only two types of light-sensitive cells designed primarily to discern light from darkness.

In a manner similar to toads, their hunting strategy requires that their prey move across the retina in order to be perceived. A frog would starve to death if surrounded only by piles of immobile food. A common toad requires only very simple information about the shape of the moving object or pattern. It interprets a horizontal black strip as prey if it is moving lengthwise. If the strip is moving in a perpendicular direction, then it is interpreted as a predator.

J.-P. Ewert **Motion perception shapes the visual world of amphibians** (2004): In: Prete F.R. (Ed.) Complex Worlds from Simpler Nervous Systems. Cambridge, MA, MIT Press, pp. 117–160

Tree frog *(Hyla arborea)*

Morays and snake eels

The eel-like morays, of which roughly 200 species are known, belong to the family of bony fish (Muraenidae) and inhabit shallow tropical and subtropical oceans. They have four nostrils – two at the tip of their snout, and two near the eyes. The nostrils are interconnected through a wrinkled system of channels that increase the inner surface through which molecules of scent may be absorbed. This gives the animals an excellent sense of smell. What is more, all four nostrils can be elongated in a tubular fashion if the occasion demands it. The anterior nostrils then extrude beyond the snout, while the posterior ones appear almost as horns between the eyes.

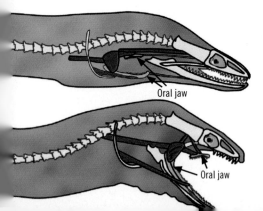

modified according to
Z. Deretsky, R. Wilson (pbroks13)

The quality of vision of these animals is not yet well understood. They appear to have large eyes with multi-row retinas, which would indicate rather good eyesight. On the other hand, morays have a reputation of being short-sighted and relying mainly on their excellent sense of smell.

Another snake-like fish

As we would assume from their snake-like figure, snake eels are fish with a relatively close relation to morays. Some species are capable of burrowing backwards, very rapidly, into the sand, relying on camouflage for further protection. Two forms of jaws exist – a regular oral jaw, and another deep inside the throat, called the pharyngeal jaw. While the nostrils of morays sit at roughly the same height as their eyes, the nostrils of snake eels extend far forward beyond the tip of their snout and culminate in a tube that is seemingly bent downwards. As with morays, we know only a little about their visual capacity. The photo below shows that their eyes may contract into a slit-shaped iris, which suggests the existence of at least two separated fovae in the retina.

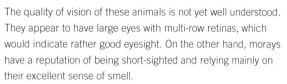

Clouded moray (*Echidna nebulosa*)

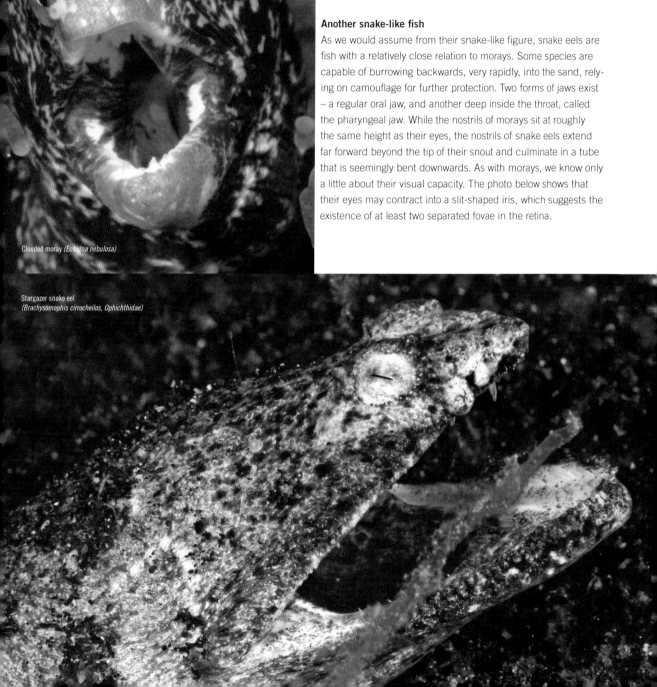

Stargazer snake eel
(*Brachysomophis cirrocheilos*, Ophichthidae)

The largest eyes among land animals

Common ostrich *(Struthio camelus)*

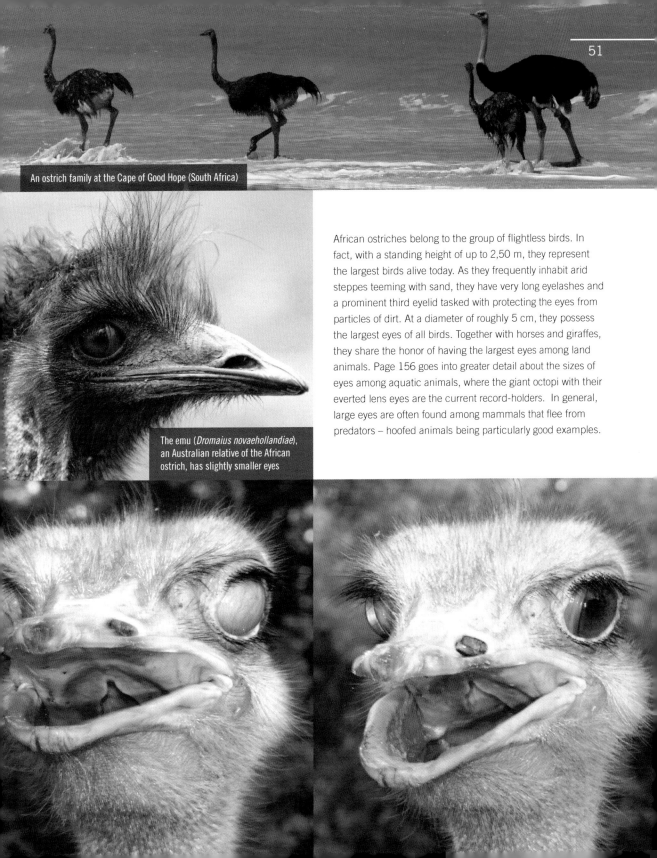

An ostrich family at the Cape of Good Hope (South Africa)

The emu (*Dromaius novaehollandiae*), an Australian relative of the African ostrich, has slightly smaller eyes

African ostriches belong to the group of flightless birds. In fact, with a standing height of up to 2,50 m, they represent the largest birds alive today. As they frequently inhabit arid steppes teeming with sand, they have very long eyelashes and a prominent third eyelid tasked with protecting the eyes from particles of dirt. At a diameter of roughly 5 cm, they possess the largest eyes of all birds. Together with horses and giraffes, they share the honor of having the largest eyes among land animals. Page 156 goes into greater detail about the sizes of eyes among aquatic animals, where the giant octopi with their everted lens eyes are the current record-holders. In general, large eyes are often found among mammals that flee from predators – hoofed animals being particularly good examples.

Colorblind?

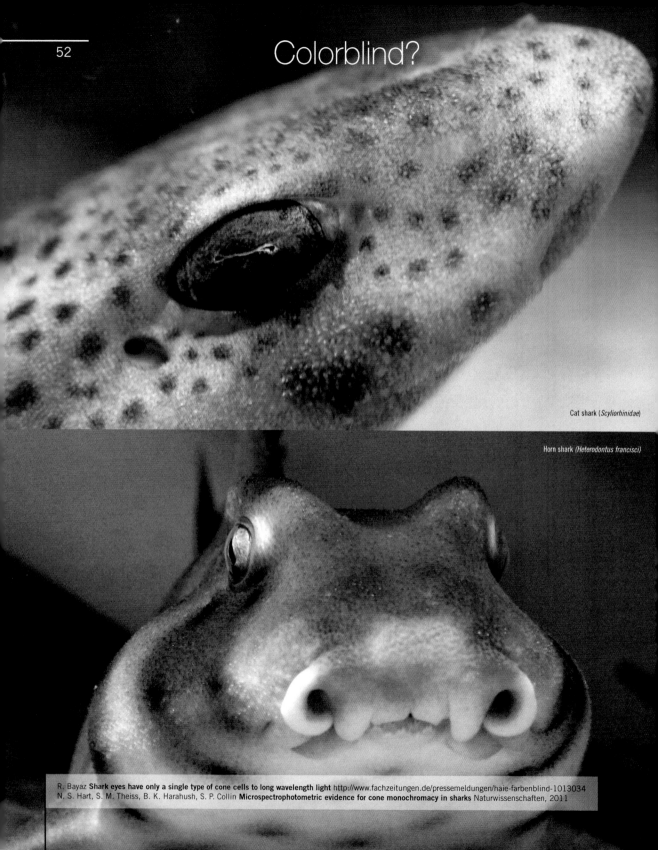

Cat shark (*Scyliorhinidae*)

Horn shark (*Heterodontus francisci*)

R. Bayaz **Shark eyes have only a single type of cone cells to long wavelength light** http://www.fachzeitungen.de/pressemeldungen/haie-farbenblind-1013034
N. S. Hart, S. M. Theiss, B. K. Harahush, S. P. Collin **Microspectrophotometric evidence for cone monochromacy in sharks** Naturwissenschaften, 2011

The eyes of sharks possess only a single type of cone cell receptive to long-wavelength light. In all probability, these animals are colorblind. Many other marine mammals – whales, dolphins, and seals among them – likewise only possess a visual cone receptive to green color. It seems that sharks and marine mammals have arrived at the same visual principle through convergent evolution.

Blacktip reef shark *(Carcharhinus melanopterus)*

Whitetip reef shark *(Triaenodon obesus)*

Cartilaginous fish

54

The structure of the eyes of rays and sharks is very much alike. This is not surprising, as both groups of animals belong to the class of cartilaginous fish. The left page shows an Indio-Pacific stingray (top photo) with an Atlantic cownose ray appearing below. The right page shows the gill opening of a blue dot ray (top photo) and the same feature of a guitarfish.

Indo-pacific stingray *(Neotrygon kublii)*

Cownose ray *(Rhinoptera bonasus)*

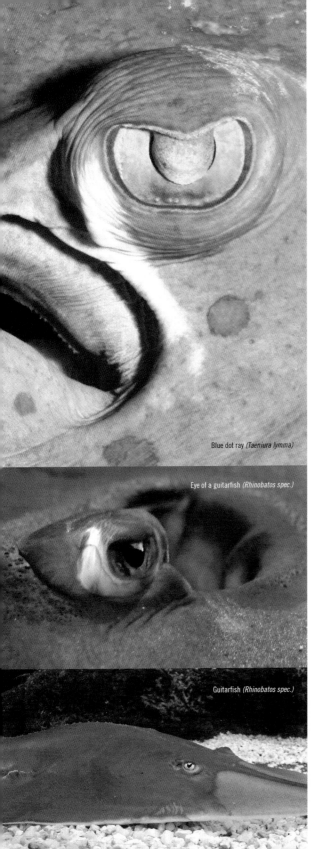

Blue dot ray *(Taeniura lymma)*

Eye of a guitarfish *(Rhinobatos spec.)*

Guitarfish *(Rhinobatos spec.)*

Rays and sharks are cartilaginous fish, as their skeleton consists of cartilage, rather than the bone substance more familiar to us. This gives them lighter weight and better properties of buoyancy, which they employ to a great advantage as free-swimming fish. Owing to their roughly conical shape, sharks are very fast swimmers. Rays, on the other hand, are almost completely flat. Their gill openings, which appear on the sides of sharks, have migrated towards their bottom. Apart from that, their pectoral fins have grown much larger, imbuing them with a wing-like grace. Rays are capable of great acceleration by propelling the water from their mouth area through their gill slits. However, their teeth are very different from those of sharks, as they feed on crabs, shellfish, and smaller fish that roam the ground surfaces. They are not as needle-sharp as shark teeth, but are instead rounded and thus used for grinding more than for cutting.

Most rays are relatively harmless. Only stingrays, with their long, whip-like tails covered in poisonous thorns, are capable of posing a serious threat to humans, although their behavior is usually defensive. Electric rays may also be dangerous in certain conditions, as they are capable of delivering substantial bolts of electricity. For this purpose, the muscle groups on the sides of their flat bodies have evolved into electrical organs. Usually, the electric tension reaches around 75-80 volt, while in some species it reaches only 25 volt. The maximum intensity discovered so far measured around 230 volts and 3-7 amperes – in special cases, it even reached as much as 30 amperes. These electrical organs are not always used defensively, but also to stun prey animals.

Like all cartilaginous fish, rays have a lens eye with a ventral retina that, by default, focuses on infinite distances. Accommodation occurs through a movement of the lens towards close-up objects (bony fish have evolved the opposite approach). Many rays further exhibit an oblique retina, producing something like glasses with gradient lenses. The dorsal retina pointing downwards is positioned at a much greater distance to the lens and is thus, by default, focused on close-up objects.

Birds and reptiles

From evolutionary biology, we know that birds are classified as feathered reptiles. In fact, birds are the only living descendants of special dinosaurs. The crowing rooster on the left and the agama on the right page exhibit a very prominent third eyelid.

Wikipedia **Nictitating membrane** http://en.wikipedia.org/wiki/Nictitating_membrane

Rooster (*Gallus domesticus*)

The third eyelid *(Plica semilunaris conjunctivae, Membrana nicitans or Palpebra tertia)* represents an additional flap of conjunctiva on the nose side of the canthus. Humans and most other primates possess this feature only rudimentarily – one notable exception being the Angwantibo. In many other vertebrates, it is transparent and able to fold itself over the eye like a pair of protective goggles.

Oriental garden lizard *(Calotes versicolor)*

Attacking the weak point

The large photos on this spread show an ant unexpectedly overwhelming a much larger hornet (in Kyoto, Japan). Its strategy was to tackle the hornet by its antenna and to keep it in a "chokehold" for half an hour and across many meters of intense confrontation, until the larger insect is finally exhausted. The hornet was then attacked and killed by other ants that came to the rescue. This situation is somewhat reminiscent of the human practice of taming large cattle by putting a large ring through their nose.

Ants sometimes hunt in large numbers, which permits them to overcome much larger prey impossible tackle on their own. Usually, however, they hunt alone. If a small prey is encountered, it is transported directly into the nest. A larger prey like a hornet propels the ant to excrete pheromones that alert other ants to the challenge.

Ants communicate primarily by means of differentiated scents. For this reason, their sense of sight is relatively unimportant – this particular species possesses only a few hundred ommatidia.

Formicine ant (*Formicinae*) grasps a social wasp by its antennae

Water buffalo (*Bubalus arnee*)

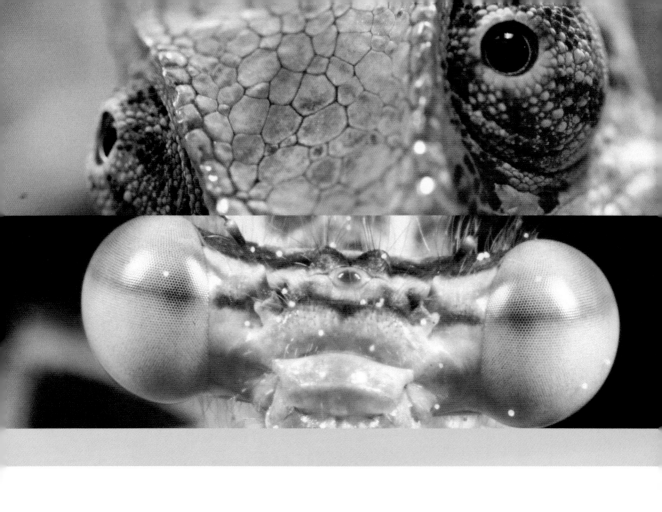

3 The world is 3D

Spatial vision

The size and positioning of many animal's eyes reveal a lot about their lifestyles. Predator eyes tend to be directed towards the front, in order to capture their prey in three dimensions. The eyes of prey animals, on the other hand, are placed to maximize the field of vision.

The world is 3D

How can light be perceived?

Light possesses several categories of stimulus important for the perception of the world. In addition to spatial and temporal dimensions, it also has qualities of wavelength and intensity, the latter being measured in the number of photons per time unit. Further categories include color (itself merely an abstraction of wavelength), contour (derived from contrasts), form, movement, and spatial depth.

Two evolutionary solutions

Essentially, light-sensitive cells are nothing more than counters of incoming photons – or rather, of changes in photon flow. An additional lens apparatus on top of a flat retina is necessary for image-based perception. The facet eyes of arthropods and the lens eyes exhibited by many other groups of animals are the two principal evolutionary solutions to this problem.

Overlapping fields of vision

Every eye possesses a field of vision. In lens eyes, it is defined as the maximum visual angle that can be perceived by its optical system. In facet eyes, it is the sum of the fields of vision of each individual facet. Spatial vision requires the work of at least two eyes, providing two distinct images for the subsequent neuronal image processing in the organism's brain. Thus, in the usual case, depth perception of the environment is accomplished by observing objects with two eyes simultaneously – a condition that requires overlapping fields of vision. If the environment is observed by two eyes from different points of view, the brain is able to assign approximate distances to each observed object. This leads to the construction of a mental model of the organism's spatial environment. In many vertebrates, this trick is accomplished without additional movements of the head.

Different eye positions

The degree to which the fields of vision of an animal's eyes overlap depends upon the positioning of the eyes on its head. The eyes of predators are usually close, and thus, their fields of vision overlap very strongly. This is especially noticeable in predatory birds – the eyes of owls, for instance, are so notoriously front-facing that they produce an almost human-like face. However, when eyes are this close together, the gain in depth perception comes at the cost of a smaller overall field of vision. This disadvantage is often overcome by more flexible movements and rotations of the head. The eyes of prey animals such as songbirds are located on the far sides of their heads for the purpose of maximizing the overall field of vision. The same observation applies to mammals, in particular hoofed animals, which also tend to have heads with the eyes located sideways. In comparison, the eyes of predatory cats are very closely aligned. Goats, horses, cows, and camels can encompass up to 330°-360° with their especially wide pupils – humans are merely capable of about 185°. Thus, prey animals gain their wide perspective by deemphasizing spatial acuity. Head posture also affects the animal's ability to see effectively into the distance. Horses have the best spatial vision when their head is raised. They are further aided by a lateral oval pupil, which projects the incident light onto a larger area on the retina.

Crossing-over of optical nerves

Depth perception is further enhanced by a crossing-over of optical nerves between the right and the left eye. In humans, the nerve pathways that cross over towards the cerebrum belong to the sensory cells of the retina closest to the nose. By means of an optical trick inherent to the eye, these particular cells receive only those rays of light that reach the retina from the direction of the temple. This trick permits image processing of the left field of vision by the right half of the brain – and vice versa. The technique of crossing-over of optical nerves is present to varying degrees among vertebrates, and among invertebrates, it only occurs in octopi. In apes and monkeys, roughly half of the optical nerve fibers overlap, which already gives these animals remarkable spatial acuity. In predators, about 75% of these fibers overlap, and in hoofed animals, this figure rises to 90%. In birds, nearly all nerve fibers cross over. Curiously, in owls, this figure is only around 60-70%. In amphibians, as in birds, nearly all nerve fibers cross over to the opposite side.

Top: The snow owl *(Bubo scandiacus)* needs forward-facing eyes to capture its prey. Center: a human eye with a closed and with an open pupil. Bottom: in jumping from branch to branch, the baboon *(Papio spec.)* must be able to estimate distances accurately.

The mantis and the mantis shrimp

Mantis shrimp *(Odontodactylus scyllarus)*

Praying mantis
(Mantis religiosa)

Eyes and pseudo-eyes of the praying mantis

Praying mantises are insects that look for prey in low vegetation. Their relatively large facet eyes are positioned to attain a large binocular field of vision, allowing them to catch prey within reach at lightning speed. When a mantis feels threatened, it opens its anterior legs, presenting the enemy a white pair of pseudo-eyes on a dark background – a pattern that bears striking resemblance to a larger predatory animal.

The complex social behavior of the mantis shrimp

Mantis shrimps inhabit the floor of tropical oceans and can often be found in caves that they dig for themselves. They grow up to 30cm in length and possess a dizzying array of legs along their elongated bodies. As in all European crabs, the anterior segment is equipped with two pairs of antennae. Upon the clown mantis shrimp's posterior pair of antennae, a large, oval protrusion is attached and is itself capable, similar to other parts of the shrimp, of exhibiting color reflections and patterns of polarization. The social behavior of these animals is quite complicated and becomes especially pronounced during territorial disputes. A clown mantis shrimp *(Odontodactylus scyllarus)* reacts quickly to intruders, communicating mainly through its wimple-like extensions, so that deadly turf battles may be avoided. These constantly moving extensions, the antennae with the oval protrusion, and the shield (telson) of the large posterior body all reflect polarized light especially well. Not surprisingly, the complex facet eyes of crabs are specialized to detect this category of light.

S. N. Patek, R. L. Caldwell **Extreme impact and cavitation forces of a biological hammer: strike forces of the peacock mantis shrimp** *Odontodactylus scyllarus* (2005) J. Exp. Biol. 208 (19): 3655-3664
S. N. Patek, W. L. Korff, R. L. Caldwell **Deadly strike mechanism of a mantis shrimp** (2004) Nature 428: 818-819

Precise estimations

Mantids belong to their own order of insects closely related to cockroaches. They are so-called ambush predators and seize their prey with lightning-fast movements. To increase their range of attack, both their prothorx and their anterior legs are strongly elongated. Their posterior pair of legs is clearly designed for walking, while the anterior pair serves to hang onto their prey. The femur and the tibia are covered in spikes, which make it easier to hold an unruly victim in place.

Giant Asian Mantis *(Hierodula patellifera)*

Central European mantids tend to wait until their prey is in the immediate range of their front legs before making a move. The conehead mantis *(Empusa pennata)* common to Mediterranean regions is even capable of catching flies in mid-air. This is only possible due to binocular vision powered by its facet eyes, through which the animal is able to estimate the position and trajectory of its prey within just a few milliseconds.

S. Rossel **Binocular vision in insects: How mantids solve the correspondence problem** 1996: Proc. Natl. Acad. Sci. USA 93: 13229–13232

European dwarf mantis
(Ameles spallanzania)

The eyes of dragonflies and damselflies

Dragonflies and damselflies belong to a primeval order of winged insects – the Odonata. Their larvae develop in the water, acting as ambush predators, while the mature forms hunt swiftly in the air aided by acute eyesight.

Pygmy damselfly *(Nehalennia speciosa)*

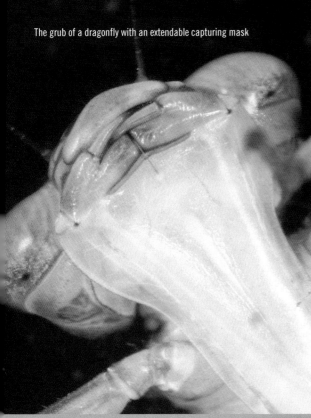

The grub of a dragonfly with an extendable capturing mask

The largest eyes among insects

The facet eyes of odonata are exceptionally large. Some species possess up to 30 000 ommatidia – a number unmatched by any other insect. The two major groups of odonata differ in the size and positioning of their eyes. While the facet eyes of dragonflies (Anisoptera) encompass nearly their whole head, even colliding dorsally in certain areas (see photo below), the eyes of damselflies (Zygoptera) are located at a distance on the sides of their heads (see left photo).

In dragonflies, the ommatidia of the doral region possess a larger diameter, which leads to a nearly parallel positioning of their optical axes. Moreover, their elongated rhabdomers are more sensitive to light and are capable of resolving images at a higher resolution. Aided by this ability, dragonflies are able to detect smaller objects in the sky – a crucial feature of their hunting strategy. Such dichotomous eyes with larger dorsal ommatidia also occur in a range of flies, although only in males. Both sexes of dragonflies have this feature, as both are tasked with hunting.

Hawker (Aeshna spec.)

Life in the sand

Weeverfish *(Trachinidae)*

Blue dot ray *(Taeniura lymma)*

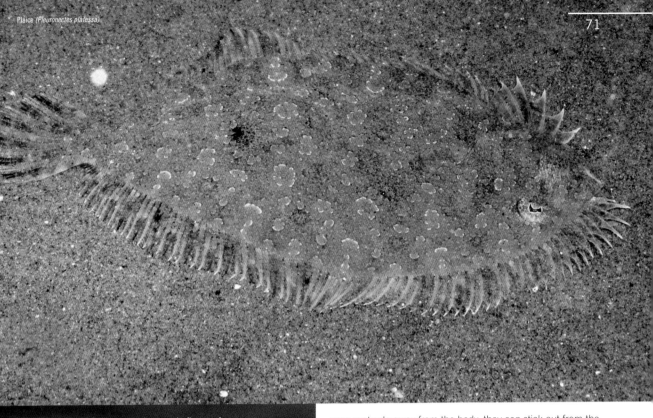

Plaice *(Pleuronectes platessa)*

Many species of fish live on the sandy floors of our seas. Some bury themselves in the sand, leaving just the head or the eyes sticking out to monitor the surroundings, while others have flattened and well-camouflaged bodies. If their eyes protrude away from the body, they can stick out from the sand while the rest of the animal is buried safely beneath it. The bluespotted ribbontail ray *(Taeniura lymma,* bottom left) provides a good example for this particular strategy.

A different approach to flattening

Flatfish like the plaice (top right) undergo a remarkable change in their bodily shape. As larvae, they still possess bilateral symmetry and swim upright in open waters. During their early development, however, one of their eyes migrates in front of the posterior fin, or through its base, towards the latter upper side, before finally coming to a halt in a bone cavity. The cranial bones and the outer nasal openings move in the same direction. Both eyes are now located on the colored upper side. From a morphological point of view, this flattening of their bodies works by completely different principles when compared to the rays. Eyes and cranial bones might either migrate to the left side (as in left-eye flounders such as the turbot), or to the right side (as in plaices and soles).

Eight legs and a large brain

A. Kühn **Über den Farbwechsel und den Farbensinn der Cephalopoden** 1950: Z. vgl. Physiol. 32: 572-598
R. T. Hanlon, J. B. Messenger **Cephalopod Behaviour** 1996: Cambridge University Press, Cambridge

Common kraken *(Octopus vulgaris)*

360 degree view through two vision slits

The eyes of octopi are located on an elevated protrusion, allowing them to have fields of vision that overlap significantly as well as a complete 360 degree view simultaneously. For this reason, their pupil is no more than a slit-shaped aperture. In fact, these animals have excellent vision, which aids them well, considering the speed at which they swim. Most octopi are colorblind, but are still able to detect polarized light. Only select species possess color vision – Octopus vulgaris, however, does not.

Rapid changes in color

Octopi are especially famous for their ability to adapt to the colors and patterns in their background almost instantly, based on a three-layer distribution of highly innervated color cells (chromatophora) on their skin. The amazing ease with which they deploy their camouflage was already known to the ancient Greeks, where the polyp was a symbol for deviousness and eclecticism.

An unusual, unusually flexible shape

The coat-like back surface of an ordinary octopus – up to 25 cm in length – features a wide opening. Its tentacles may reach a length of up to a meter and are covered by two rows of suction cups. Small ocelli (point eyes) reside between the eyes and on the base of the lateral arms.

Memory and learning

Many experiments have demonstrated the extraordinary intelligence of octopi. They quickly learn to unscrew closed jars containing food, and are able to memorize complex paths through narrow passageways when they move freely without their shell.

Common kraken *(Octopus vulgaris)*

Harvestmen and octopuses

Harvestman *(Opilio spec.)*

Cellar spider *(Pholcus phalangioides)*

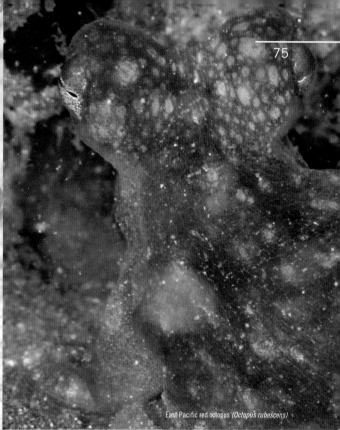
East Pacific red octopus *(Octopus rubescens)*

Mimic octopus *(Thaumoctopus mimicus)*

Notice the resemblance between the harvestman (left photo) and the octopi in these photos. Like octopi, harvestmen possesses two lens eyes on an elevated protrusion, which allow for 360-degree vision. Being arachnids, they are also equipped with eight legs, and are further aided by elongated chelicera which serve as their trophi.

The analogy between harvestmen and octopi can be extended even further. Harvestmen have a compact, egg-shaped body, as their two basic body parts (prosoma and opisthosoma) have become completely fused. They only possess one pair of median eyes, and the laterally positioned remainders of what were once facet eyes have become completely reduced. Finally, the most well-known species of harvestmen are notorious for their exceptionally long legs.

Other arachnids that sometimes roam European appartments are often mistaken for harvestmen due to their long legs. The top left image shows one such example, the cellar spider *(Pholcus phalangioides)* – in this case a female, grasping and guarding its bundle of eggs with its trophi.

Double image processing

Double image processing

Chameleons (Chamaeleonidae) encompass more than 160 species, most of which can be found in Madagascar (where the whole clade most likely originated), Africa, the Middle East, and Southern Europe. Among lizards, they belong to the family of agamids. Their build, diet, and defensive postures are well-adapted for a life in trees. Species often differ substantially between each other, and furthermore, their form often depends on gender and age.

Two independent, movable eyes

Chameleons have lens eyes that can be moved independently and that protrude from their cavity in an effort to increase the field of vision. As with the rest of their skin, their eyelids are covered in scales and have grown together with parts of the eyeball. The only aperture resides in front of the lens in the manner of a pinhole camera and serves to increase contour sharpness.

Distance estimation through measurement of accommodation

All reptiles focus on varying distances by deforming their lens in a precise manner (accommodation) across a range of over 45 diopters. It is presumed that the brain of the chameleon is able to correlate the contraction of the eye muscles with the distance of the prey being observed. This would allow the animal to estimate its distance very precisely, and it would explain the accuracy with which the chameleon launches its tongue towards its prey.

Few blind spots and enormous sharpness

A chameleon is able to see 90° vertically and 180° horizontally. When both eyes are considered, its field of vision reaches 342°. The only blind spot, measuring 18°, resides on the back, directly above its head. As in humans, visual sharpness is obtained through the cornea. Since the

Veiled chameleon
(Chamaeleo calyptratus)

Parson's chameleon *(Calumma parsonii)*

whole eye is covered by the eyelid, which leaves only a small aperture for incoming light, the optical qualities resemble those of a pinhole camera, which leads to a sharper image. Chameleons are able to see sharply and clearly at distances of up to one kilometer. Their focussing speed can reach 60 diopters per second – roughly four times what the human eye is capable of. A hunting chameleon uses both eyes simultaneously and independently to scan its surroundings, with moving targets being examined by both eyes as the fields of vision partially overlap. It is presumed that the visual information coming from an eye is processed by the chameleon's brain independently and separately from the other eye.

A. Lustig, K-K. Hadas, K. Gadi **Threat perception in the chameleon (Chamaeleo chameleon): evidence for lateralized eye use** (2012) Animal Cognition 15: 609–621
A. Herrel, J. J. Meyers, P. Aerts, K. C. Nishikawa **The mechanics of prey prehension in chameleons** (2000): J. Exp. Biol. 203: 3255-3263
P. Necas **Chameleons: Nature's Hidden Jewels** (1999): Krieger Publishing Company
F. Le Berre, B. Le Berre, D. Richard **The Chameleon Handbook. Barron's Educational Series** (2009) 3rd Edition. ISBN 0764141422.

4 The limits of clarity

Fine tuning of retina and optics

Any good eye requires high-quality optics and/or a high-quality retina. Birds of prey gain the most from improving their retina, while some crabs are remarkably successful at growing high-fidelity optics in each of their facets.

The limits of clarity

Adaptations of eyes

As a general rule, animals tend to have eyes as good as their manner of living in their particular environment requires. When it comes to maximizing luminance efficacy, different adaptations become necessary. Nocturnal animals and those of the deep sea must make the most from a small number of photons. Water poses a problem, as it absorbs a lot of light energy – especially of the long-wavelength variety. As animals descend further and further below the water surface, short-wavelength blue light starts to predominate (see the image on the right). Light cannot penetrate much deeper than 1 km, even when the water is completely clear.

For some organisms, discerning light from darkness is all that is required. In such cases, a mere light-sensitive patch composed of a few sensory cells may suffice. When the direction of light becomes important, a sensory epithelium – either flat or recessed into the epidermis – must evolve. A system of lenses is required to accomplish a truly sharp image on this type of retina. This necessitates that the retina resides within the focal plane with respect to parallel light rays emanating from distant objects. For objects nearby, however, the plane of sharpness moves outside of the eye. This can be compensated through accommodation. Either the whole lens is moved forwards and backwards (as in fish and reptiles), or the curvature of the lens is changed by muscle contractions (as in mammals).

Varifocal glasses

Many hoofed animals (such as horses) have evolved another method for discerning different planes of sharpness. Their retina grows in an oblique manner such that its upper half is further away from the lens than the lower half. The effect thus achieved is similar to the optical properties of varifocal glasses and allows the animal to focus on grazing with the upper part of the retina whilst the lower part remains focused on the horizon.

The visual sharpness depends on many constructional properties of the eye. A mapping system onto a flat retina is necessary, but not sufficient. A further neuronal system must be in place to process the incoming image. For invertebrates, this system begins through a series of axons, outbound from each sensory cell, which are then connected to further processing centers, either in the brain or in the so-called *Lobus opticus*. In vertebrates, the axons of both light sensory cell types (rod and cone cells) are already interconnected near the retina. In that sense, some image processing is already taking place before the signals reach the brain. Increased visual acuity is usually accomplished by a multiplication of receptors in the retina as evident in birds of prey. In arthropods, the ommatidia are multiplied instead, as in the case of dragonflies.

Increasing luminance efficacy

As photographers know, there are several ways to increase luminance efficacy. Larger lenses capture more light, but this comes at the price of greater distortion, which further decreases sharpness. Facet eyes solve this problem by collapsing the separation between individual ommatidia, permitting light to be gathered by multiple ommatidia simultaneously (consider the superposition eyes of nocturnal insects). Higher crabs have evolved another, even more ingenious method. Their crystal cones are no longer used as converging lenses, but rather as mirror reflection mechanisms.

Giant barrel eyes

Deep-sea fish inhabit a world almost completely bereft of light. This poses a problem for predators, who must, after all, locate their prey precisely. In order to utilize what little light is available, many species have evolved barrel eyes. These are exceptionally large, elongated, and possess an enormous spherical lens with a long focal length. Their side walls are transparent, allowing light to fall onto the so-called auxiliary retina, while most of it passes through the large lens onto the main retina. The retina itself is almost exclusively composed of rods sensitive to short-wavelength (blue) light. What is more, the rods are extremely tightly packed – sometimes even on three superimposed layers. Here, it can be said that evolution has achieved a genuine optimum uniquely tailored to the deep-sea environment.

Shellfish with mirror eyes

Pilgrim mussels *(Pecten)* possess parabolic mirrors in their eyes, of which they have approximately 60. Here, light falls onto the hemispherical background of their eyes which is covered by a reflective layer – the *tapetum lucidum*. The double retina is hit twice. The second retina is significantly more sensitive and appears to aid nocturnal vision.

A number of nocturnal animals possess reflective layers on the background of their eyes. The eyes of cats are especially famous for their reflective quality. The *tapetum lucidum* can thus be considered as a low-light amplifier.

Hammerhead shark (*Sphyrnidae*)

Telescope eyes

This photograph shows that sharks can definitely focus on targets in front of their noses

Bonnethead shark *(Sphyrna tiburo)*

The family of sharks known as hammerhead sharks (Sphyrnidae) is named and famed for their widened heads (also called cephalofoils). It encompasses eight species split into two genera, mainly differentiated by the shape and width of their heads – some of which have grown so wide as to truly deserve the name "hammer". The largest species are the great hammerheads *(Sphyrna mokarran)*, reaching up to 5.5-6.1 meters in length, while the smallest, the scalloped bonnethead *(Sphyrna corona)*, only grows up to a meter in length. Their eyes reside at the far end of their cephalofoils, are round or nearly round in shape, and possess an inner third eyelid. The reason behind the cephalofoil shape is not well understood. The shark's field of perception is definitely enlarged by the shape of its head, though it is further assumed that the widened head aids the animal during maneuvering – in fact, "cephalofoil" loosely translates as "head foil", derived from the aviation term "airfoil". The cephalofoil's function is much closer to that of the canard wing – another concept from aviation, ai-

D. M. McComb, T. C. Tricas, S. M. Kajiura **Enhanced visual fields in hammerhead sharks** (2009): J. Exp. Biol. 212: 4010–4018
B. R. Brown **Modeling an electrosensory landscape: behavioural and morphological optimization in elasmobranch prey capture** (2002): J. Exp. Biol. 205: 999–1007

...med at improving an airplane's control over its vertical trajectory through additional airfoils at its nose. One particular application of this technique is the improvement in maneuverability and lift when navigating narrow curves. From the shark's perspective, a generous cephalofoil allows for the pectoral fins to be smaller than those of other sharks.

The eyes and the enlarged nasal cavities lie at the far ends of the widened head, leading to a significantly larger field of sensory perception. The same applies to the ampullae of Lorenzini at the front of the cephalofoil, which are not only able to sense electrical impulses of potential prey, but also the Earth's magnetic field – a capacity that aids them well during long migrations, to which some species of hammerhead sharks are prone. As their heads grow wider, their fields of perception increase, as do the overlapping fields of vision, allowing for an improved three-dimensional (binocular) sense of sight.

High resolution

Wikipedia **Bird vision** http://en.wikipedia.org/wiki/Bird_vision
G. R. Martin **Form and function in the optical structure of bird eyes** (1994): in
M. N. O. Davies, P. R. Green **Perception and motor control in birds** Springer, S. 5-34
R. Shlaer **An eagle's eye: quality of the retinal image** (1972): Science 176 (4037): 920-922
H. P. Ziegler, H.-J. Bischof (Hrsg.) **Vision, Brain, and Behavior in Birds: A comparative review** (1993): MIT Press
O. Güntürkün **Sensory Physiology: Vision** (2000): In G. C. Whittow (Hrsg.): Sturkie's Avian Physiology, Orlando: Academic Press, S. 1-19
O. Güntürkün **Structure and functions of the eye** (1998): in G. C. Whittow (Hrsg.): Sturkie's Avian Physiology (5. Edition), Academic Press, S. 1-18

Golden eagle
(Aquila chrysaetos)

Eagle owl (Bubo bubo)

For birds, sight is the most essential of all senses. The eyes of some species have grown so large that they can hardly be moved. In great eagles, the distance between the lens and the retina measures roughly 2,5 cm – more than in humans – which produces larger and sharper images on the retina.

Reptile eyes capable of seeing ultraviolet light

In terms of general structure, bird eyes are equivalent to those of their reptilian ancestors. Their lenses undergo greater deformation than those of mammals. Most birds have four receptor cells on their retina (pigeons have five), ranging from red to ultraviolet. The ability to detect UV light allows birds of prey to detect traces of mouse urine (which reflect UV light), while fruit-eating birds use UV light to assess the ripeness of their food. The density of receptors per retina and the sheer number of nerve connections to the brain is usually higher than in mammals. What is more, almost half of all receptor cones are so-called "double cones".

A better detection of slow and fast movements

A higher refresh rate allows the animal to resolve fast movements more accurately. While humans tend to perceive motions faster than 50 Hz as continuous, this threshold often reaches 100 Hz in birds. Without such acuity of motion, sparrow hawks that chase songbirds through branch-wood would quickly collide with their entangled surroundings. Birds are also able to resolve very slow motion – for instance, the sun, moon, and stars appear to them as moving objects and aid them during migrations.

Unsurpassable birds of prey

Relative to the sizes of their bodies, birds of prey have especially large eyes, possessing up to 65,000 receptor cells per square millimeter (almost double the amount of humans). In experiments, it was shown that eagles are able to tell a mouse apart from a stone of similar size and color from a height of 600 m. American falcons detect 2 mm long insects from the top of a tree measuring 18 m. Most birds of prey are further equipped with a double fovea, which permits them to focus simultaneously on large parts of their visual fields.

Of crows and vultures

On the right: A baby crow has fallen out of its nest. The iris of its eyes is still blue (unpigmented). It will take weeks before the final dark colors are reached. On the left: a griffon vulture invades the airspace of a crow's nest. An adult crow keeps attacking the vulture in an attempt to chase it away. Vultures have excellent eyesight and are able to detect carrion from a distance of many miles.

Griffon vulture *(Gyps fulvus)*

Young crow *(Corvus corone)*

Round or slit-shaped pupils?

Cougar (*Puma concolor*)

The mammal family Felidae encompasses big cats (tigers, lions, jaguars, etc.) and small cats (wild cats, domestic cats, ocelots, etc.). The eyes of all cats are relatively large when compared to the sizes of their skulls. Their pupil apertures are highly flexible. In well-lit environments, they are slit-shaped in small cats, while other types of cats exhibit pupils of a small and circular shape in these situations. Only in darkness do these pupils open widely. Some species of cats possess multifocal lenses, allowing for increased sharpness – especially in low-light situations when their pupils need to be fully open. Domestic cats have multifocal lenses combined with a slit-shaped pupil. Lions and tigers are only equipped with monofocal lenses.

T. Malmström, R. H. H. Kröger **Pupil shapes and lens optics in the eyes of terrestrial vertebrates** (2005): J. Exp. Biol. 209: 18-25

Tiger *(Panthera tigris)*

Lynx *(Lynx lynx)*

Cheetah *(Acinonyx jubatus)*

Lion *(Felis leo)*

Evergreen State College **Anatomical Schematic of a Cat's Eye** http://archives.evergreen.edu/webpages/curricular/2011-2012/m2o1112/web/cats.html
E. Guenther, Zrenner, Eberhart **The Spectral Sensitivity of Dark- and Light-adapted Cat Retinal Ganglion Cells**
(April 1993) Journal of Neuroscience 13 (4): 1543–1550. PMID 8463834   Domestic cat *(Felis catus)*

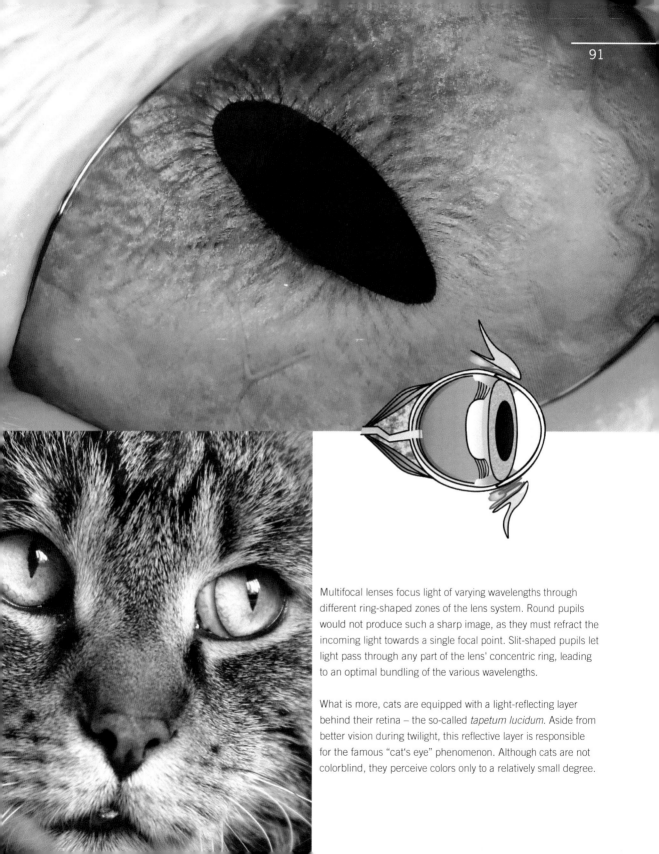

Multifocal lenses focus light of varying wavelengths through different ring-shaped zones of the lens system. Round pupils would not produce such a sharp image, as they must refract the incoming light towards a single focal point. Slit-shaped pupils let light pass through any part of the lens' concentric ring, leading to an optimal bundling of the various wavelengths.

What is more, cats are equipped with a light-reflecting layer behind their retina – the so-called *tapetum lucidum*. Aside from better vision during twilight, this reflective layer is responsible for the famous "cat's eye" phenomenon. Although cats are not colorblind, they perceive colors only to a relatively small degree.

Shark vision

Sandtiger shark *(Carcharias taurus)*

Great white shark *(Carcharodon carcharias)*

White sharks are often found in low-visibility waters, where sharp imaging is only possible for a few meters. Thus, they have to depend on other senses, such as their lateral organs and ampullae of Lorenzini.

Sharks are visual animals

Sharks mainly depend on their magnificently evolved eyes, despite other excellent sense organs at their disposal (such as their perception of electricity). The pupils of shark eyes are either circular or oval, and are aligned either diagonally or vertically. In contrast to other bony fish, sharks are able to change the size of their pupils through their iris. By default, a shark's lens is focused on far distances. However, it can be moved with the aid of muscles – as the photo on the left page shows, where the animal definitely appears to be focused on the camera lens nearby. The white shark (seen on all remaining pictures) occasionally lifts its head above the water to inspect its surroundings. Thus, it stands to reason that its eye is capable of adapting to the varying refractive properties below and above the water surface.

Color blind, but highly light-sensitive

The retina of sharks is covered with numerous receptor cones and rods. Unlike their close cousins – the rays and the ratfish – sharks are colorblind, since all of their receptor cones are of the same type. Due to their unusually large amount of receptors, and as a corollary their ability to detect small differences in luminance contrasts, sharks excel at black-and-white underwater vision in poor visibility conditions. In order to boost luminance efficacy, sharks have a *tapetum lucidum* behind their retina. It consists of thousands of polygonal plates covered by a layer of guanine crystals capable of strong reflections akin to the silver of a mirror. The light passing through the retina is thus reflected, increasing the eye's luminosity – similar to how cat's eyes appear in the dark. In order to avoid damaging the eyes during confrontations with prey, some species are equipped with a third eyelid, which moves in front of the eye during the last phase of attack. Other species are able to roll their eyes towards the back to protect them from damage (see bottom right photo).

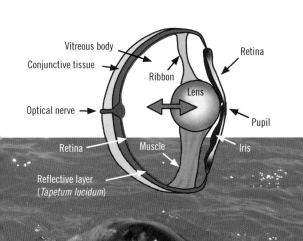

Great white shark *(Carcharodon carcharias)*

N. S. Hart, S. M. Theiss, B. C. Harahush, S. P. Collin
Microspectrophotometric evidence for cone monochromacy in sharks
(2011): Naturwissenschaften 98 (3): 193-201.

Two yellow spots on the retina

Octopuses do not belong to vertebrates, but they have similar eyes

The octopi (belonging the subclass Coleoidea) are a subgroup of cephalopods. Despite being neither fish nor vertebrate, the more highly evolved species among them possess complex lens eyes which resemble the eyes of most mammals (including humans) in their basic construction. They possess an eyeball with an anterior transparent cornea, an additional spherical lens, and a complex retina at the back of the eye, from which all axons are bundled into an ophthalmic nerve that runs towards the brain.

Squid *(Loligo spec.)*

Sepia *(Sepiida)*

An octopus eye has no blind spot
The light-sensitive receptor cells in octopus eyes are oriented towards incident light (everted) – contrary to the eyes of all vertebrates, where the cells point away from the inciding light (inverted). For octopi, this difference has remarkable consequences for the neuronal processing of visual stimuli. Light receptor cells transfer the electrical signals toward the brain through axons at the lower end of the sensory cell. If this cell is oriented away from the light – as in vertebrates – then the light must first pass through the axon layer and the cell body before reaching the apical part of the cell, where the actual perception takes place. In octopus eyes, this detour is not necessary. Light that passes through the lens hits the apical parts of the light-sensitive cells directly. In the eyes of mammals, this path is considerably obscured by further layers of nerve cells where a first processing of the light information is already taking place, and through which light must pass in order to reach the photoreceptors. For this reason, the construction of vertebrate eyes makes it impossible to absorb light at the point where the nerve cells are routed towards the brain – the famous "blind spot". Octopuses require no such compromise.

Differences in embryonic development
Aside from further differences in the construction of light sensory cells, the eyes of octopi differ from those of mammals mainly in their embryonic development. As in most other invertebrates, the whole octopus eye emerges entirely from the embryo's outer cell layer (ectoderm) through complex eversion processes. The vertebrate eye, on the other hand, is produced from the embryo's different germ layers. The whole vertebrate retina, with its many layers of nerve cells, primarily represents the eversion of tissue from which the brain is produced at a later stage. The cornea and the lens, on the other hand, emerge from the outer blastodermic layer – the ectoderm. A similar mechanism produces these features in octopi. The eyes of octopi and vertebrates are remarkable evolutionary solutions that have converged on similar techniques from two completely different hereditary standpoints. These historical preconditions are responsible for the considerable differences that persist in their details.

Excellent vision, though rarely in color
Octopi possess over 100,000 receptor cells per square millimeter. Aided by their lens system and a curved, large-surface retina, their sense of vision is excellent despite being blind to color. Only firefly squid possess three photopigments (like humans and many species of insects) in the microvillus membranes of their sensory cells. However, many other species are able to detect polarized light – presumably to better detect reflecting light on silver-colored prey fish.

Spatial vision in vertebrates ...
Binocular (and thus spatial) vision is usually achieved by moving both eyes forward with parallel visual axes, producing the impression of a "face" as in owls, apes, and humans.

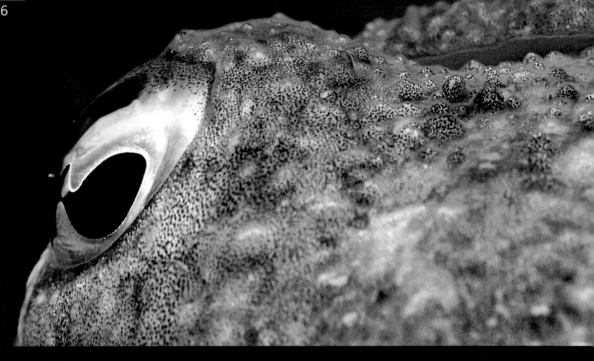

... and invertebrates

Octopi accomplish a large field of view and simultaneous depth perception by one of two separate ways. The familiar cuttlefish, for instance, are ambush predators in shallow waters and whilst the position of their eyes at their sides of their bodies would seemingly indicate a lack of spatial vision, their pupils are w-shaped, which means that the light is being focussed on two separate parts on the retina. In fact, two regions exist on the retina for sharp vision – one directed toward the back, and another toward the front. This permits the animal to have simultaneous sharp vision toward its front, and its surrounding environment.

The eyes of kraken (Octopoda), on the other hand, lie on a strongly elevated protrusion, giving these animals overlapping fields of view as well as a "panoramic perspective". The pupils of kraken are merely slit-shaped apertures, which does not stop them from using their excellent vision for very fast swimming maneuvers.

Squid *(Loligo spec.)*

Squid

Common cuttlefish *(Sepia officinalis)*

Unbeatable

As mentioned on page 39, the eyes of mantis shrimp are incredibly well developed – far superior to those of other crabs and far more complex than the eyes of most other animals. In some species, these precision instruments may consist of up to 10,000 ommatidia. The central strip is not only able to analyze 100,000 colors, but also ultraviolet and polarized light.

Mantis shrimp *(Odontodactylus spec.)*

J. Marshall et al. **Behavioural evidence for polarisation vision in stomatopods reveals a potential channel for communication** (1999): Current Biology 9 (14): 755-758
T.-H. Chiou et al. **Circular Polarization Vision in a Stomatopod Crustacean** (2008): Current Biology. 18, S. 429
J. Marshall, M. F. Land, C. A. King, T. W. Cronin **The compound eyes of mantis shrimps (Crustacea, Hoplocarida, Stomatopoda)** in **Compound eye structure: The detection of polarized light** (1991): Phil. Trans. R. Soc. Lond. B 334: 57-84.

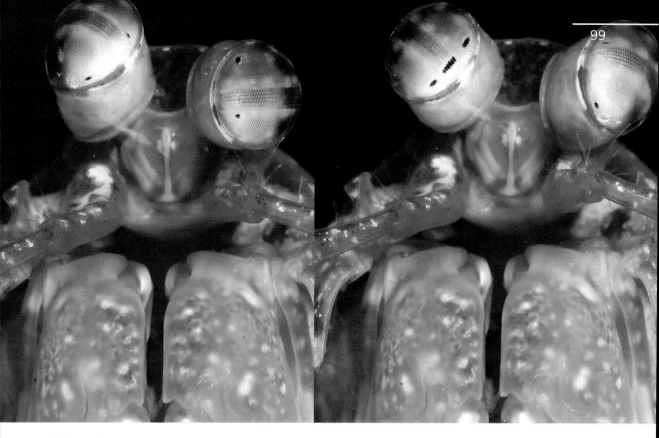

Spatial vision with only one eye

The eyes of a mantis shrimp (the photos depict an *Odontodactylus scyllarus*) share much of their structure with the apposition eyes of other crabs. Their central part, consisting of six ommatidia rows, divide the eyes into upper and lower hemispheres. The optical axes of the six rows are exactly parallel and directed forwards, while the first rows of both hemispheres cross over each other, being slightly inclined towards the center.

The outer rows run in parallel to the central rows, and the further they are from the middle, the more inclined they become towards the outer side. This construction, unique among animals, allows the crab to see a triple image with a single eye – thus, spatial vision becomes possible. While the upper and lower hemispheres serve the detection of shape and motion, the highly complex sensor in the middle, with its six rows of ommatidia, is responsible for the detection of colors and polarization. This sensor's angle of view is not especially large, spanning only about 10-15 degrees. Due to its independently flexible eyes, the crab is able to use one eye to assess an object's shape, while using the other to detect its colors and polarization.

Seeing and interpreting polarized light

Depending on the sun's location on the firmament, the polarization pattern changes as the light hits the atmosphere. Determining the sun's position on a completely overcast sky is among the skills of animals capable of detecting the orientation pattern of waves. Honeybees, among other species, use this skill to their advantage. A hitherto unknown pattern recognition was also discovered in mantis shrimp *(Stomatopoda)*, which were found to be able to detect circular polarized light. Such light can be imagined as a sort of spiral, where the polarization plane rotates in the light's propagation direction. Crabs are even capable of distinguishing between light polarized in a left and right direction. This form of perception plays a role during mating behavior and remains completely hidden from other animals. Artificial polarization filters accomplish the same task during DVD or CD playback, and also in some types of cameras. These manufactured filters, however, are only able to work with one color of light, while the eyes of mantis shrimp excel almost perfectly across their whole visible spectrum – from near-ultraviolet to infrared. So far, similar reflection patterns have only been observed in Glorious Jewel Scarab *(Chrysina gloriosa)*.

Are eight eyes better than two?

From facet eyes to lens eyes
Terrestrial arachnids descend from ancestors living in the sea. It is thought that the facet eyes of these ancient animals regressed as they made their move onto land in the Silurian. At first, only five lenses remained, which themselves were composed of multiple fused ommatidia. In later generations this number was reduced to only three small lens eyes. However, these were accompanied by a larger pair of median eyes. This led the majority of primitive arachnids – the scorpions – to have a total of eight eyes. The pair of median eyes quickly evolved into their primary visual organ, while the lateral lenses mainly serve the differentiation of light from darkness and the detection of prey in motion (see page 105 on the bottom right).

Primary and secondary eyes
Real spiders (Araneae) have retained this set of eyes. They are thus equipped with one pair of large median eyes, which have, as in other arachnids, evolved into their most potent visual organs. In real spiders, they are often called primary or anterior median eyes, while the other six (three on each anterior side) are called secondary eyes. Depending on the group of spider, however, these smaller eyes may be very differently distributed – thus, many families of spiders are recognizable merely based on the positioning of these eyes. One pair of secondary eyes is often located next to the median eyes and is thus referred to as the front lateral eyes.

Jumping spider: Inverse retina and a large focal length
From the point of view of their construction, the primary and secondary eyes differ in that the latter are often inverse, while the secondary eyes are always everse. The inverse retina is always coupled to a *tapetum lucidum*, allowing light to be reflected. Jumping spiders possess no less than four (tetrachromatic) receptor cell types, which also happen to be very numerous. The strongly enlarged and forward-facing main eyes are equipped with big vitreous bodies, which produce large focal lengths.

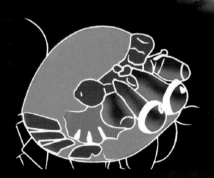

Jumping spider from Thailand *(Siler semiglaucus)*

F. G. Barth **A Spider's World: Senses and Behavior** (2002): Springer-Verlag, Berlin-Heidelberg-New York, p. 394
D. Neuhofer, R. Machan, A. Schmid **Visual perception of motion in a hunting spider** (2009): J. Exp. Biol. 212(17): 2819-23
T. Nagata, et al. **Depth Perception from Image Defocus in a Jumping Spider** (2012): Science 335 (no. 6067): 469-471

Four retinal layers for the estimation of color and distance

The lens is focused on the four underlying retinal layers depending on the wavelength of light. The lowest two layers are receptive only to green color, however, the image is only sharp on the lower one. This difference in sharpness between the two retinas is used by the animal to estimate the distance of an object.

Spatial vision by a movement of the retina

Through a rotation of their anterior bodies, jumping spiders are able to quickly and noticeably change the direction in which they look. In addition, the retina can be moved by three pairs of muscles such that the spider can extend the visual field of its main eyes, leading to an overlap with the auxiliary eyes – thus allowing for spatial vision. A sharp color image of prey or partner can be seen at distances exceeding 10 cm. These spiders are able to estimate distances by comparing degrees of sharpness with respect to the focal plane – a strategy that we, as humans, also employ. However, spiders are able to see much more with eight eyes than with just two.

A cooperation of eyes

There exists something like a division of labor between these eight eyes. Aided by six muscles, the main eye is able to scan and analyze an object at the focus point of the remaining eyes.

For many other spider species, vision is not so important

Not all spiders have such a sophisticated pair of primary eyes. Depending on their lifestyle – whether diurnal or nocturnal, and whether sitting on cobwebs or on flower blossoms – these eight eyes may vary significantly in size and positioning. The lower image shows a crab spider that lives as an ambush predator on blossoms, preying on insects that feast upon the flower's nectar. Their eyes play only a subordinate role in this endeavor, which explains their small size. Tarantulas likewise have only small eyes, as they are able to localize their prey by detecting vibrations on their elaborately woven webs before stunning their unfortunate victims with their poisonous claws.

Goldenrod crab spider
(*Misumena vatia*)

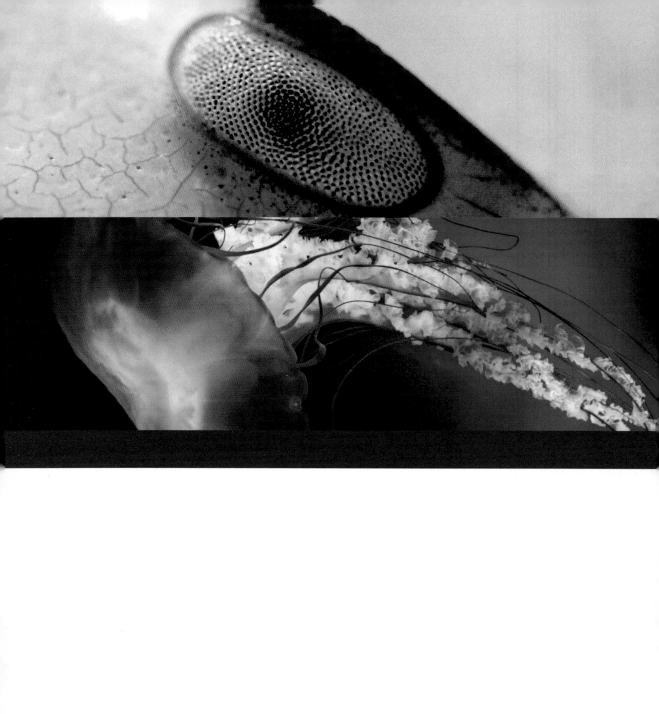

5 Simple or simplified?

When seeing clearly is not required

The simplest eye consists of just one sensory and one pigment cell. Another precursor of the lens eye – the pinhole camera eye – shows that "simple" does not always mean "primitive". Ancestors of millipedes millipedes had more complex eyes, which have become simplified.

Simple or simplified?
When seeing clearly is not required

The path of light to the brain

Light-sensing cells contain light-sensitive molecules that convert these impulses into electrical energy. However, they are only able to identify changes in photon flux. To detect more information, a retina is needed that transfers electrical impulses through axons to the brain. This further requires a light-harvesting apparatus that projects a realistic image onto the plane of receptors – usually, this is accomplished through a pinhole camera or a lens system. In fact, the kingdom of animals abounds in transitional forms between simple photoreceptors and highly complex lens eyes. Biologists designate eyes as "simple" if they are merely able to differentiate between light and darkness, with only a small capacity to detect direction or shape. Sponges, flatworms, many species of annelids, and nearly all sessile molluscs possess a so-called "diffuse sense of light". Singular photoreceptors are distributed across their skin, which help them identify spots

(Ophiocomidae) possess a large number of tiny lenses made of calcite on all five of their arms. Their focal point is directed precisely onto the underlying nerve fibers. This makes it plausible to

assume that they are able to capture rather good images of their surroundings.

The prototypical eye

Marine annelids of the genus *Branchiostoma* also possess very simple eyes, each consisting of no more than a single light sensory cell covered by a lens cell. The former includes vertically stacked membranes made of cilia that contain the light-absorbing pigment rhodopsin. This enables the animal to detect the direction of incident light. In an evolutionary context, these types of eyes are called "prototypical eyes" (see also the chapter on "pax and homology"). Most species with prototypical eyes underwent a grouping of receptor cells into a retina under amplified selection pressures towards better perception of their environ-

of optimal brightness, and avoid others that are too dark or too light. Some species still retain these receptors in addition to their "real" eyes and use them to gather information about the color and structure of the ground below. Molluscs, whose photoreceptors are sensitive to color, use this ability to great effect. As they lie buried in the sand, they pull back in a near-instant reflex if a nearby fish happens to casts a shadow upon them. It was long thought that echinoderms (like starfish or brittle stars) possess no eyes, or at best, only a diffuse sense of light. However, some species of brittle stars that live in tropical waters

ments. At first, the retina was lowered into a pit (consider the pit eyes of marine snails), which gradually became so deep that any inciding light had to pass through a small aperture. The pinhole camera eyes of the marine snail *Haliotis spec.* or the primitive octopus *Nautilus spec.* are good examples of this approach. Finally, the pit was covered by a transparent layer, which led to the repeated convergent evolution of lenses, from which the many varieties of complex lens eyes evolved that we find today in the animal kingdom. The prominent group of vertebrates also began its evolutionary journey with mere photoreceptors. The

J. Aizenberg, D. A. Muller, J. L. Grazul, D. R. Hamann **Direct Fabrication of Large Micropatterned Single Crystals** (2003): Science 229 (no. 5610): 1205-1208

lancelets of the genus *Branchiostoma* – a group of originally headless vertebrates – do not have eyes as such, but their spinal cord is covered in light-sensitive receptors. An unpaired black spot – the frontal eye? – is especially prominent. Recent findings have shown that there are similar opsins at work as in the photoreceptors of most other vertebrates. This has confirmed the old conclusion that the light-sensitive cells of lancelets are precursors to more modern vertebrate eyes. As modified nerve cells, such sensory cells are the precursors to the cones and rods found in more sophisticated eyes. In other animals, light-sensitive cells descended from epidermis cells.

Secondary simplification through depletion and reduction

Animals whose ancestors have had good lens eyes can still shrink or reduce them if their habitat has changed to be underground (naked mole rats and moles) or in dark caves (olms). The same applies to cave-dwelling insects, where transitions between barely functioning and completely reduced eyes are often found. Olms living in Slovenian and Croatian caves are usually blind, however, there is a population living above ground possessing small eyes and dark-pigmented skin. Naked mole rats are also assumed to be completely blind, however, it was discovered that they have a relatively well-developed retina, lacking only a dioptric apparatus. Why these animals would need a retina at all is still unclear.

The body plan of arthropods

Arthropods possess from four to eight point eyes (nauplius eyes, frontal ocelli) and a facet eye. All original members of the arthropod group, such as marine arachnids *(Limulus)*, extinct trilo-

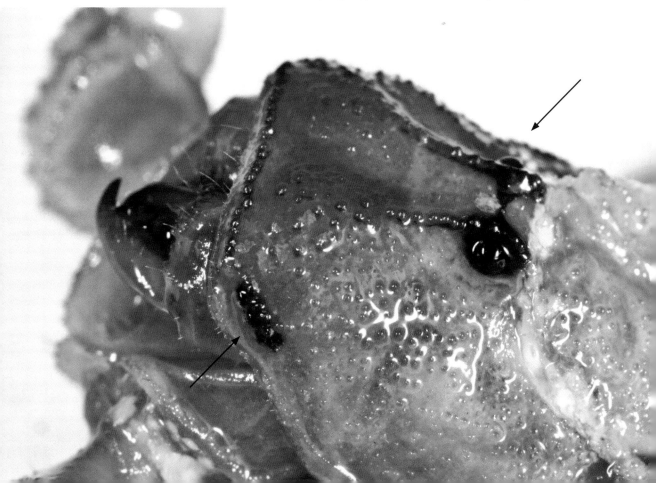

Sessile feather duster worm *(Sabellina)*

bites, crabs and insects, centipedes and millipedes *(Myriapoda)* descended from ancestors with facet eyes. Some species like the house centipede *Scutigera* – the only recent members of the Myriapoda – still retain their facet eyes, even though it is unclear whether they have kept the ancient form or simply "reinvented" them more recently.

The eyes of arachnids

The distant ancestors of modern arachnids lived mostly in the sea. Today, they are represented by two groups – one that never left the water (horseshoe crabs, *Limulidae*) and another that conquered land as the Silurian gave way to the Devonian (more than 420 million years ago). The latter animals evolved from sea scorpions with facet eyes to all modern terrestrial arachnids, including the land-dwelling real scorpions of today. None of them have facet eyes, but merely retain remainders in form of individual lens eyes and a pair of median eyes. Many scorpions have up to five isolated lenses at the sides of their anterior bodies in addition to their more prominent nauplius eyes. Real spiders, like cross spiders or wolf spiders have a total of eight such lenses. Two of them are the aforementioned median eyes, while the remaining six, arranged in groups of threes on each anterior side of their bodies, are the remainders of ancestral facet eyes.

Arachnids are usually nocturnal

It is reasoned that as the early arachnids made their move onto land, they were forced to inhabit a niche of "night and fog" due to missing protection against evaporation in their cuticula, which in turn led to a strong change of their facet eyes. The majority of modern arachnids are nocturnal. However, some of them became diurnal predators that have further adapted their sense of sight. The jumping spiders are especially famous for having perfect lens eyes for their two median eyes (see page 100). This is a good example of an important evolutionary principle – that evolution tends to optimize certain functions by way of gradually building upon precursor structures.

From larval eyes to high-performance eyes

Similar simplifications of facet eyes can also be found in the larvae of holometabolic insects (see page 114). Adult butterflies, for instance, are equipped with large facet eyes, but their larvae only have six strongly modified lateral lenses called larval eyes (stemmata). Some insect larvae have developed these stemmata, which were rather simple at first, into high-performance lens eyes in a secondary feat of evolution. In larvae of the diving beetle species *Acilius* (grooved diving beetle) and *Thermonectus*, two of their six stemmata develop into strongly enlarged bifocal lenses, whose separated focal planes coincide with two different retinas within the same eye.

Millipede *(Julidae)*

Coral scallop *(Pedum spondyloideum, Pectinidae)*

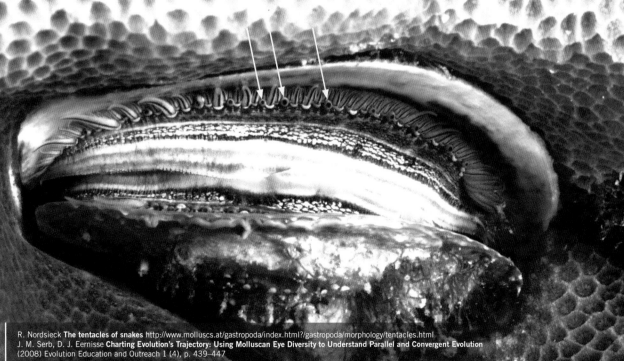

R. Nordsieck **The tentacles of snakes** http://www.molluscs.at/gastropoda/index.html?/gastropoda/morphology/tentacles.html
J. M. Serb, D. J. Eernisse **Charting Evolution's Trajectory: Using Molluscan Eye Diversity to Understand Parallel and Convergent Evolution**
(2008) Evolution Education and Outreach 1 (4), p. 439–447

A very primordial eye

D.-E. Nilsson **Eye evolution and its functional basis** Visual Neuroscience 30, p. 5-20

Emperor nautilus *(Nautilus pompilius)*

Living fossils with pinhole camera eyes

The ancestors of cephalopods possessed a simple pinhole camera eye without a lens that we may still encounter in six species of so-called "living fossils", one of which is the nautili (nautilus and allonautilus) that inhabit the tropical reefs of the Western Pacific and some parts of the Indian ocean.

In earlier times, these primordial lifeforms had a much richer and diverse variety of relatives distributed across Earth's waters. The first nautilids emerged towards the end of the Cambrian – roughly 500 million years ago – and flourished during the Ordovician period in particular. The extinct ammonites, which, like today's nautilids, also featured rolled-up limestone shells, counted among their close relatives. Their gas-filled chambers provided them with lift, enabling them to navigate slowly through the waters.

The opening of a pinhole camera eye acts as a narrow aperture, focussing the inciding light onto the retina. Due to the absence of a lens, no refraction of light needs to occur – instead, the eye cavity fills up with water. The visual clarity thus accomplished is only very limited, but since these animals move very slowly, this is not of much importance.

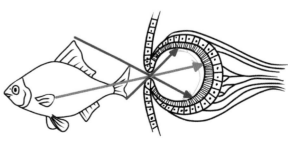

"Pro forma" eyes

Animals tend to have the kinds of eyes that their particular lifestyles require. It may be tautological, but it is still usually true to observe that very simple eyes indicate a lifestyle that does not require a particular sense of vision. Whether an animal's vision represents a primordial or a secondary feature in its ancestral lineage is a different, more interesting question. In the case of the millipedes (myriapoda), this remains an open subject for debate. Earlier opinion used to hold that the myriapoda are the closest cousins of insects, whose ancestors lived in the water and later adapted for life on land by evolving, among other features, a tracheal system which allowed them to breathe air. Their taxonomical group, the Tracheata, owes its name to this characteristic. The ancestors of tracheata and crabs were marine animals with facet eyes. Since the facet eyes of crabs and insects are remarkably similar, it is argued that the eyes of myriapoda must be secondarily evolved complex eyes, as they do not have facet eyes made of ommatidia. Like terrestrial arachnids, the ancestors of myriapoda also adapted their facet eyes into simple collections of lenses. In fact, molecular analysis has shown that myriapods are much older than originally thought, and that crabs are actually the closest cousins of insects. This

Millipede
(Alcimobolus domingensis)

Rough woodlouse *(Porcellio scaber)*

might indicate that the eyes of myriapoda originated as simple collections of lenses similar to ommatidia. However the situation might have actually been – and it is still being debated among scientists – it is clear that these eyes are relatively simple compared to the facet eyes of crabs and insects. By comparison, it is much easier to come to a conclusion about the eyes of ispods, which come from a line of crabs that conquered land. Despite breathing through gills, they have found an elegant solution that enables them to inhabit land permanently. They retained their gills on their ventral sides within pouches filled with water, breathing as if submerged in the ocean. Like millipedes, they have become nocturnal, feeding on the waste products of plants. Their large facet eyes – originally telescopic – shrunk and moved closer to their head. It seems beyond doubt that these represent a secondary feat of evolution.

H. F. Paulus **Phylogeny of the Myriapoda – Crustacea – Insecta: a new attempt using photoreceptor structure** (2000): J. Zool. Syst. Evol. Research 38: 189-208
H. Miyazawa et al. **Molecular phylogeny of Myriapoda provides insights into evolutionary patterns of the mode in post-embryonic development** (2013): SCIENTIFIC REPORTS, 4 : 4127, DOI: 10.1038/srep04127

A diversity of miniature eyes

A small animal with a cyclops-like eye
Tiny shrimp, measuring only a few millimeters in length, can often be noticed in our ponds by their "jumping" motions as they move their large antennae back and forth in order to be propelled through the water. Their bodies are enclosed in a double shell. Miniscule appendages on their insides are used to filter microorganisms from the water. Their unpaired facet eyes consist of 22 traditional ommatidia with a crystal cone containing eight retinula cells each.

A sphere of facets, merged in a secondary step
Like all crabs, these animals originally had two separated eyes that merged in a secondary evolutionary step, producing a sphere of facets. These are connected to a series of muscles that permanently induce small twitching motions, akin to the subtle motions that constantly occur in our own eyes. The purpose of these motions is to keep the optical image constantly on the retina so that it may be processed more accurately. All such movements, including those of the whole organism and those of the object being observed, lead to a perpetually changing light stimulus on the retina. It is assumed that the tiny crab uses these changes for better orientation.

The first highly evolved eyes in the animal kingdom
Box jellyfish belong to a class of marine cnidarians, which count among the most primordial multi-celled organisms. They inhabit a fixed location as polyps before assuming a solitary, free-floating existence as medusae. Their name stems from the cuboid shape of their shields during their medusa stage. Most famously, these animals possess stinging cells that are released upon touch. The projectiles are

Water flea (*Daphnia pulex*)

B. J. Frost **Eye movements in *Daphnia pulex* (De Geer)** (1975): J. Exp. Biol. 62: 175-187
A. Garm, M. M. Coates, R. Gad, J. Seymour, D.-E. Nilsson **The lens eyes of the box jelly fish *Tripedalia cystophora* and *Chiropsalmus sp.* are slow and color-blind** (2007): J. Comp. Physiol. A, 193: 547-557
D.-E. Nilsson, L. Gislén, M. M. Coates, C. Skogh, A. Garm **Advanced optics in a jellyfish eye** (2005): Nature 435: 201-205
H. Suga, P. Tschopp, D. F. Graziussi, M. Stierwald, V. Schmid, W. J. Gehring **Flexibly deployed Pax genes in eye development at the early evolution of animals demonstrated by studies on a hydrozoan jellyfish** (2010): Proc. Natl. Acad. Sci. USA 107 (32): 14263-8

often strongly venomous and can cause serious skin reactions in many species. The venom of box jellyfish is especially strong, with some species of sea wasps (of the genus chironex) having the capacity to kill humans. Box jellyfish are free-floating predators and were the first species of the animal kingdom to develop high-quality eyes.

A complete panoramic view
Due to their tetraradial body plan (possessing no anterior and posterior), they have developed several different types of eyes that sit on four outer radii, giving them a complete panoramic perspective. A total of 24 eyes are located on the so-called rhopalia – among these are two lens eyes, two slit-shaped pigment eyes, and two pit-shaped pigment eyes (auxiliary eyes). A vestibular organ is further found at the base of each rhopalium. Interestingly enough, the lens eyes – lacking color vision – are oriented towards the inner part of their shield. In controlled experiments, medusae were quite adept at using them to avoiding dark objects.

A box jellyfish of the species tripedalia is pictured on this page. Tripedalia play a big role in how biologists discuss the evolution of eyes in the animal kingdom (see chapter 7 for more on this topic), as they already possess genes for the development of eyes that also take part in eye formation of vertebrates.

Young medusae of the box jellyfish *(Tripedalia cystophora)*

Where are the eyes?

Eyes and mouthparts at the front

In determining the front and the back side of a particularly unusual animal, we normally try to look for its eyes. The left picture shows the grub of a swallowtail butterfly (Papilio machaon), whose eyes are rather hard to find (see the arrow). This task becomes even more difficult in caterpillars of vapourers (Orgyia antiqua, top right) – in this particular case, the animal's anterior points to the left side. It might be helpful to look for mouthparts, which are usually located near the animal's eyes (see bottom right). However, it may not always be easy to differentiate mandibles from mere protruding bristles which the animal uses for protection.

Caterpillar of a swallowtail *(Papilio machaon)*

H. F. Paulus **Evolutionswege zum Larvalauge der Insekten - Ein Modell für die Entstehung und Ableitung der ozellären Lateralaugen der Myriapoda von Facettenaugen** (1986): Zool.Jb.Syst. 113 (3): 353-371.

Caterpillar of a vapourer *(Orgyia antiqua)*

The remainders of a larger facet eye

Six barely discernable lenses are distributed across the side of the animal's head. As in spiders, these larval eyes, or so-called stemmata, represent the remainders of an earlier, larger facet eye. The larvae of all insects that undergo a complete metamorphosis (with a pupal stage) possess only such stemmata. The evolutionary paths from ancestral facet eyes to singular lenses were complicated and must have occurred repeatedly and independently. However, the pressures that drove these changes must always have been the same. Larvae live in concealed locations in which they only need very simple eyes to survive. A butterfly caterpillar's most important concern is where its leaves are. It evades predators passively through camouflage or by secreting toxic substances from its body.

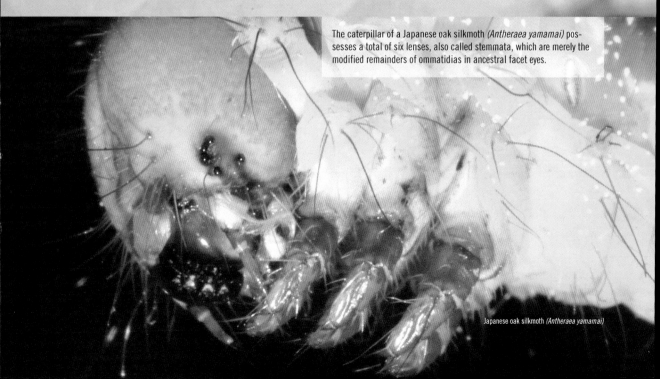

The caterpillar of a Japanese oak silkmoth *(Antheraea yamamai)* possesses a total of six lenses, also called stemmata, which are merely the modified remainders of ommatidias in ancestral facet eyes.

Japanese oak silkmoth *(Antheraea yamamai)*

Eight eyes do not make a visually oriented animal

K. P. Mueller, T. Labhart **Polarizing optics in a spider eye** (2010): J. Comp. Physiol. A 196(5): 335-48

Yellow sac spider *(Cheiracanthium)*

Spiders are world champions of mechanoreception

The sensory world of spiders is focused primarily on their incredibly subtle detection of vibrations and air turbulences. Their four pairs of eyes only serve the detection of motion, with no particular emphasis being put on sharp imaging. The jumping spiders mentioned on an earlier page are an exception. This page shows a yellow sac spider *(Cheiracanthium punctorium)* – one of only two species of European spiders dangerous to humans. The genus itself can be found throughout the world and encompasses around 200 species mostly indigenous to Southern Europe and the southern part of Central Europe. As in many other species of spiders, the females are tasked with guarding the bundle of eggs, mostly in tall grass under a bell-like web.

Claw sizes and dangers to humans

Humans are usually bit by these spiders after touching or even wiping across their web. At a body length of 15mm, this spider, being equipped with long, needle-sharp cheliceras, is strong enough to inject venom into our skin. Such a bite can be painful, but is usually not any more dangerous than the sting of a wasp. While all spiders are venomous, only a few are able to penetrate human skin with their claws. The relatives of the yellow sac spider (genus *Drassodes*) are quite adept at nocturnal orientation, using the smaller eyes behind the two main eyes to detect polarized light in the sky and to find their way back to their daytime hideout.

The simple eyes of insects

Additional light-sensing organs

In addition to their facet eyes, insects may possess up to three additional light-sensing organs located between their antennae. These median eyes of arthropods are known by several names. In arachnids, they are called median ocelli, in web spiders, they are called primary eyes, and in insects, they are called frontal ocelli.

Is something wrong with their retinas?

The light-sensing organs of insects tend to have well-developed, bulged lenses, with a construction which resembles that of small lens eyes with a cup-shaped retina. Curiously enough, the focus of these lenses is located behind the retina. One might, therefore, conclude that the lenses are not suitable for the detection of sharp images. However, more precise measurements in dragonflies have shown that several ingenious lens features compensate for this apparent astigmatism. What is more, the retina is placed precisely as to allow sharp imaging of different wavelengths.

What is the use of insect frontal ocelli?

Despite research, very little is currently known about the functioning of these eyes. Some evidence suggests the existence of unspecified organs for the measurement of brightness that regulate the sensitivity of the facet eyes. In other species, such as in dragonflies or grasshoppers, these organs could aid the detection of the horizon during flight. Crickets, on the other hand, are equipped with nothing more than a bright, translucent window in the cuticula, without any of the properties attributed to lenses.

Not all insects have frontal ocelli

Frontal ocelli already occur in primitive insects (springtails, bristletails) and are broadly common among winged insects. However, they are absent in most beetles, and in the larvae of holometabolic insects, centipedes, and millipedes. Nonetheless, these insects make up for their lack of frontal ocelli by laterally positioned stemmata, which have developed from primordial facet eyes.

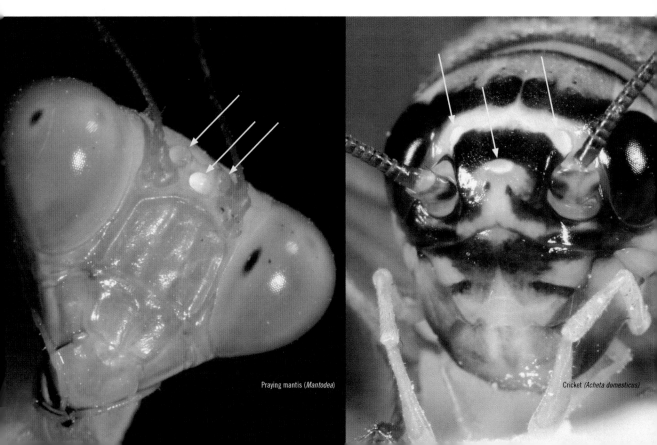

Praying mantis (*Mantodea*)

Cricket (*Acheta domesticus*)

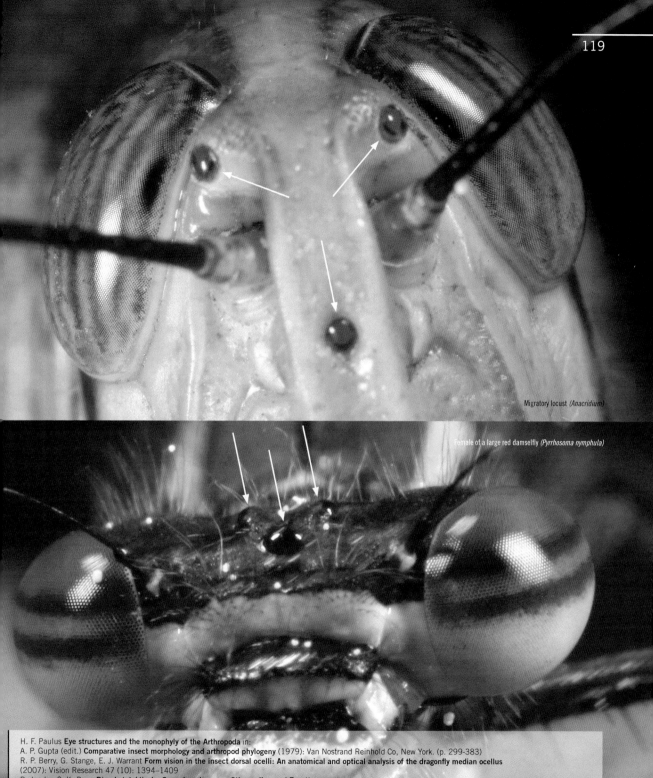

Migratory locust *(Anacridium)*

Female of a large red damselfly *(Pyrrhosoma nymphula)*

H. F. Paulus **Eye structures and the monophyly of the Arthropoda** in:
A. P. Gupta (edit.) **Comparative insect morphology and arthropod phylogeny** (1979): Van Nostrand Reinhold Co, New York. (p. 299-383)
R. P. Berry, G. Stange, E. J. Warrant **Form vision in the insect dorsal ocelli: An anatomical and optical analysis of the dragonfly median ocellus** (2007): Vision Research 47 (10): 1394–1409
R. Jander, C. K. Barry **Die phototaktische Gegenkopplung von Stirnocellen und Facettenaugen in der Phototropotaxis der Heuschrecken und Grillen (Saltatoptera:** *Locusta migratoria* **und** *Gryllus bimaculatus***)** (1968): Z. Vergl. Physiol. 57: 432-458

The square facets of higher crabs

On page 40, we have shown that the facet eyes of certain species of crabs, such as crayfish, lobsters or shrimp, exhibit square patterns instead of the usual hexagonal patters. From a geometrical point of view, these eyes are remarkable:

1. A thin, square prism is able to reflect a light ray almost like a mirror

A light ray can pass through a perpendicular, mirrored, hollow, and square prism in many ways. The number and sequence of reflections determines the direction in which it eventually makes its escape. The case shown in the diagram is especially interesting and occurs with varying frequency depending on the angle of incidence and the height of the prism. In the top view (on the right), the ray exits the prism in parallel to the direction of incidence. The directional vector of the light ray changes its algebraic sign for the first two of its components, but not for the third. If the diameter of the prism is very small, as in the eye of the crab, the following approximation becomes true: The light ray acts as if it was being reflected in an axis-parallel plane (perpendicular to the top view of the inciding ray).

2. The sphere-shaped arrangement of facets focuses a part of the reflecting rays

If hundreds or thousands of prisms are positioned on a sphere, then these act as mirrors perpendicular to the sphere's surface. All parallel rays that are reflected in the right manner are thus focused towards one point on the directional straight line that passes through the sphere center, roughly half of its radius away from it. The "image point" of the ray direction thus lies on a concentric sphere with a halved radius.

3. All possible directions of rays produce an upright, luminous image on a sphere

Following only geometrical considerations, if the angle of incidence of the incoming rays is varied (analogous to varying the position of the distant points from which these rays originate), then the image point thus produced occurs at the intersection of the "main viewing ray" and the spherical retina.

More images at: www.uni-ak.ac.at/evolution

Common prawn (Palaemon serratus)

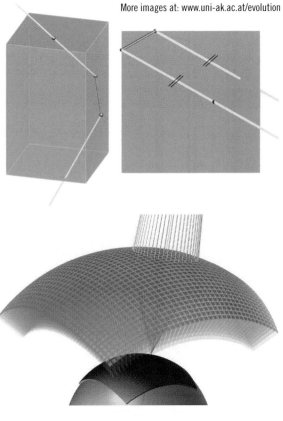

Fossils eyes also deserve a closer look ...

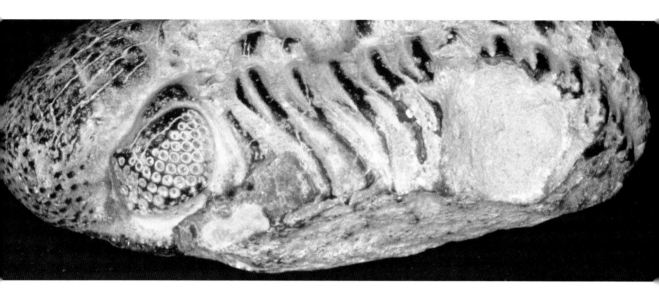

Trilobites, an early subgroup of ancestral, marine arthropods, were already equipped with two facet eyes that differed in several important details from the facet eyes of crabs and insects – insofar as their fossilized state even permits morphological analysis. By the arrangement of their ommatidia, we can differentiate two or sometimes even three types of trilobite eyes.

Type 1: Hexagonal (holochroal) facet eyes

The large majority of species possessed holochroal eyes – "regular" eyes with a hexagonal pattern of facets, consisting of a set of densely packed, convex lenses comprised from calcite that were completely covered by a single corneal layer of the same material. The singular lenses were hexagonal in shape and varied in their number from a single one up to more than 15,000 crystals. They functioned more or less like the facet eyes of modern insects, with the only significant difference being that the eyes completely consisted of clear calcite crystals.

Typ 2: Schizochroal visual apparatus

These remarkable organs are unique to the order of phacopida. Schizochroal eyes were made from a few up to 700 large, thick, and round lenses that stood more or less on their own. Each individual lens resided in a conical or cylindrical socket and was morphologically as well as optically separated from its adjacent lenses. The uniqueness of the schizochroal eye lied in its construction. The individual lenses were nearly spherical and relatively large, focusing light towards a single focal point. This led to the usual problem of spherical aberration, due to the different distances that light rays have to travel depending on their angle of incidence and the material of the lens. No single focal point existed, and therefore, no truly clear image could have formed. A sort of shell resided at the base of each individual lens that differed in structure and was also clearly demarcated from the area above it. Thus, schizochroal lenses are doublets. The calcite crystal absorbed magnesium atoms, which led to a change of its refractive index and to a correction of the spherical aberrations. The schizochroal Phacops lens was able to detect larger portions of its surroundings – presumably with considerable sharpness and clarity. Although its cellular construction was not preserved, several imprints of cells could be reconstructed. The construction of the eyes resembled that of the horseshoe crab (on page 36).

M. S. Lee, J. B. Jago, D. C. García-Bellido, G. D. Edgecombe, J. G. Gehling, J. R. Paterson
Modern optics in exceptionally preserved eyes of Early Cambrian arthropods from Australia (2011): Nature 474 (7353): 631-634
B. Schoenemann, E. N. K. Clarkson **Discovery of some 400 million year-old sensory structures in the compound eyes of trilobites** (2013): Scientific Reports 3, Article 1429, S. 1-5 (http://www.nature.com/srep/2013/130314/srep01429/pdf/srep01429.pdf)

Flying foxes

Bats represent an order of ancestral mammals whose anterior legs developed into flight organs. A so-called patagium (flying membrane) emerged alongside their bodies and between all of their fingers (excepting their thumbs), reaching right up to their hind legs and even their short tail. On a global level, bats are subdivided into two large groups – the microbats *(Microchiroptera)* and the flying foxes *(Megachiroptera)*. While the former is found on all continents, the latter group lives

Flying fox *(Pteropus spec.)*

only in Africa, and in the area between Southeast Asia and North Australia. Microbats primarily feed on insects. Their South American varieties also feed on blood, fruit, and flower blossoms. Flying foxes feed in a similar way, excluding blood from their diet. The nocturnal microbats use ultrasound to communicate with each other. What is more, they are so precisely guided by their power of echolocation that they are able to "see" their environment through their auditory sense alone. Hence, their small eyes play only a subordinate part.

Flying foxes are active at twilight and thus visually oriented, which accounts for their larger eyes. Due to their winged front legs and their much larger build, they are barely able to walk. Instead, they use their legs to hang upside-down from tree branches in large numbers during daytime. In fact, they perform most important activities in this position, including feeding, lactating their young, and reproducing.

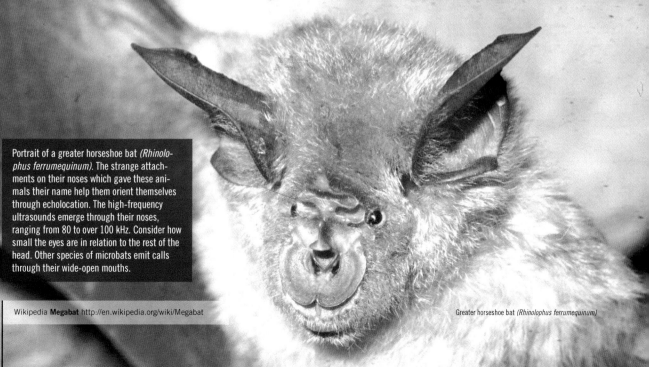

Portrait of a greater horseshoe bat *(Rhinolophus ferrumequinum)*. The strange attachments on their noses which gave these animals their name help them orient themselves through echolocation. The high-frequency ultrasounds emerge through their noses, ranging from 80 to over 100 kHz. Consider how small the eyes are in relation to the rest of the head. Other species of microbats emit calls through their wide-open mouths.

Wikipedia **Megabat** http://en.wikipedia.org/wiki/Megabat

Greater horseshoe bat *(Rhinolophus ferrumequinum)*

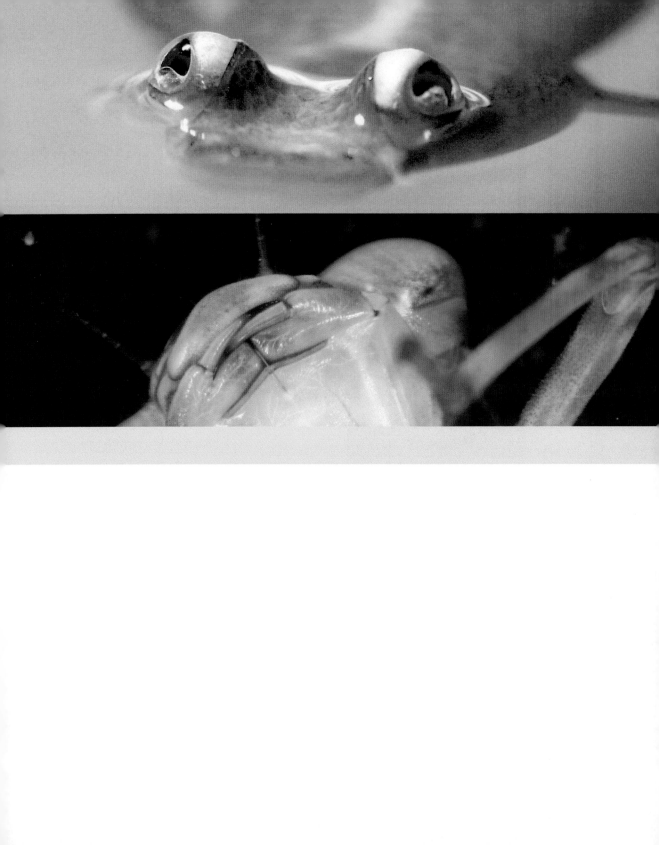

6 Above and below the water

Different requirements for each medium

The eyes of water animals are different from those of their land-based cousins. Above the water, the outer cornea is more important to clear vision than the lens. Below the surface, the bulk of refraction is done by the spherical lens.

Above and below the water

The invention of lenses represented a considerable advance in image perception. For instance, in octopi with only pinhole camera eyes, the image projected onto a retina is noticeably darker than in octopuses with lens eyes. Organisms that dwell in water mostly exhibit spherical lenses, yet still manage to attain a sharp image on the retina due to a decreasing refractive index from the lens center towards its outer regions (also called an inhomogenous lens). This reduces refractive distortions considerably and mostly shortens the overlong focal distance, enabling a clear picture on the retina. Terrestrial animals are able to employ their cornea as a refractive medium, which accounts for about two-thirds of the total refraction occurring in their eyes. The lens is mostly responsible for adjustment of focal length (accommodation). Animals who must see well in and outside of water have to evolve special adaptations. In such cases, the cornea is often used as a mere glass panel, with all refractive duties falling on the lens. Diving birds are able to change their lenses under water by up to 80 diopters. Most fish also have a planar cornea, whose refractive index barely differs from that of water. As these fish jump out of the water, they become very shortsighted for the duration of the flight due to the refractive properties of their cornea. This poses a problem for fish who have to look through water and air on a regular basis.

The four-eyed fish *(Anablebs anablebs)* that lives in South American brackish water feeds mostly on insects sitting on the water surface. Its bulging eyes have migrated upwards, and their upper halves frequently peek above the water surface. In order to facilitate vision through both mediums, the cornea is split into two parts – one for vision above, and another for vision below the water. The upper half, dedicated to vision through air, is only slightly curved, while the lower half refracts more strongly.

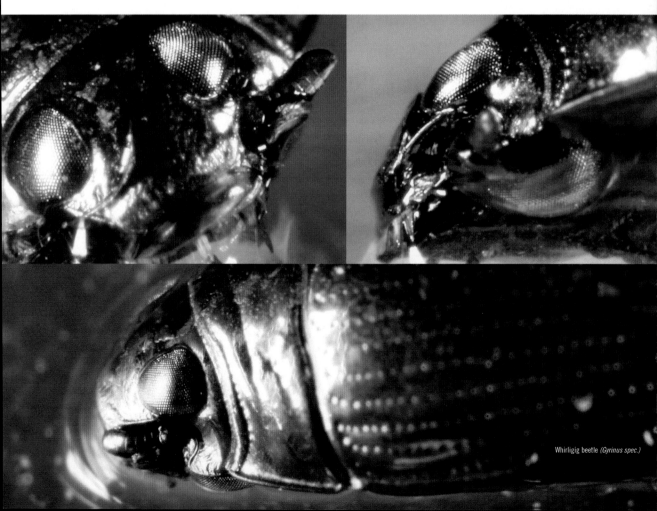

Whirligig beetle *(Gyrinus spec.)*

The lens within the eye is oval in shape. Its flat side is directed towards the upper cornea, while its strongly curved side points towards the submerged cornea. For this reason, the retina is also separated into two parts. Its ventral side serves the imaging of objects above the water and possesses twice the amount of cones as its dorsal counterpart. From the outside, they look completely separated, giving the appearance of four distinct eyes.

The family of whirligig beetles (Gyrinidae), which counts among the water-scavenger beetles, swims on the surface of the water, keeping part of their heads submerged. Their facet eyes have separated to produce four distinct eyes, able to see above and below water. The cornea of the upper facet eye is more strongly curved than that of the ventral eye.

These beetles feed on insects that have fallen into the water. In order to locate their victims, they use circling swimming motions to produce water waves that carry a bow wave at their front, which reflects objects that swim in the water – such as prey insects – whose reflected patterns the beetles then detect with their antennae. In principle, this resembles echolocation on the water surface.

Diving water birds are faced with the same problem of seeing well in both mediums. As penguins or the great crested grebes dive below the water, their corneas lose part of their refractive capacity. These are already rather flat, and barely contribute to the total refraction in the medium of air. Under water, however, their lenses are strongly curved, contributing drastically to the refraction of the optical apparatus.

Another interesting problem arises for flying predators looking for prey below the water surface. A gray heron, attempting to capture a fish under water, must consider the optical illusion caused by the refraction of light at the water surface, by which the fish appears in a seemingly different location. This shift of position depends on the depth of the fish in the water, and on the distance of the bird's head from the water surface. Herons are able to compensate for these distortions perfectly, and usually succeed at snatching their prey from the water. The same is true for the plunge-diving birds, such as the halcyon, the booby that lives close to the sea, and the South American brown pelican.

The blue-footed booby pictured on this page is about to attempt a nosedive. It must, therefore, assess how to hit the water surface in order to catch its prey.

Its story is continued on the next page ...

Blue-footed booby *(Sula nebouxii)*

A simple eye model above and below the water

Blue-footed booby *(Sula nebouxii)*

On the left, a hunting blue-footed booby penetrates the water surface. The bottom picture shows possibly the last image seen by the fish. A third eyelid is instantly superimposed, causing a change in refraction under water.

A curved cornea suffices on land

Let us imagine a model eye, spherical or otherwise, that is filled completely with a water-like substance. From one side, it should be connected to the outer air through a transparent sphere cap (red). The transparent outer layer of the sphere cap (model cornea) should have a realistic refractive index with respect to the inner liquid. According to the law of refraction, light that falls incident from distant points through the sphere cap is refracted towards the perpendicular axis. If we then use a pupil-like aperture to limit the incident light to the light rays near the optical axis, we will observe the light to be redirected towards a single focal point F. If this point is located at the boundary of the model eye, then we have already found an optical system that is able to focus sharply on distant points. If the sphere cap has only a small radius (as in the eyes of mice, for instance), then the "land eye" thus produced is already quite effective, as most imaged points would be relatively distant compared to the small size of the eye.

A spherical lens is required underwater

The following problem arises underwater: the liquid in our model eye may have a slightly higher refractive index than water, but it can never reach the value 4/3, which occurs at the transition between air and water. This causes the focal point to be very far from the pupil – except if the cornea is very strongly curved. Therefore, this problem can be solved by adding a second liquid (or another transparent material) with a slightly higher refractive index, in the overall shape of a sphere. This moves the total focal point of the optical system towards the boundary of the eye, as in the terrestrial model.

The lens eyes produced by evolution almost always exhibit a built-in "auxiliary lens", which has a radius of curvature that can be changed by muscle contractions. This enables the focusing of points at different distances, even very short ones ("accommodation"). Nevertheless, there exists a quantitative difference between land and water animals. The sea lion in the lower pictures, for instance, has good vision under water, but is very short-sighted above it. In the medium of air, the outer sphere cap refracts the light so strongly that its focal point is located inside the eye.

Galapagos sea lion (Zalophus wollebaeki)

Adaptation to vision under water

The eyes of seals
Seals have relatively large eyes with excellent underwater vision. Compared to the size of their bodies, their eyeballs are larger than those of humans by a factor of one third. They are adapted to the lower light intensity under water by having a greater amount of light-sensitive rods on their retina. However, seals are blind to color, as only cones or multiple types of rods grant color vision.

Tapetum lucidum
The spectrum of light sensitivity varies depending on the environment. For example, elephant seals have eyes which are most sensitive to light in the blue spectrum. In common seals, which mostly inhabit coastal waters, this spectrum is shifted in the direction of green. Their *tapetum lucidum*, a reflecting layer behind the retina, represents another adaptation to low-light situations. It improves light efficacy by reflecting rays that pass through the retina.

Seals require the spherical lenses of fish
As previously mentioned, a cornea has almost the same refractive index as water. To compensate, the lens of seals is, as in fish, nearly spherical in shape, leading to much greater refractive capabilities. While fish accommodate by pulling the lens forwards or backwards, seals are able to deform it in a mammalian way. They still remain very short-sighted on land.

Multifocal lenses and flexible pupils
A seal's lens is multifocal. Through accommodation, it can increase its depth of focus if the pupil is expanded. This combination of morphological properties and the functional attributes of the retina provides seals with relatively good vision at depths of up to 400-500 m. They are further able to tolerate the high light intensities that often occur in their terrestrial environments – such as on ice-covered surfaces illuminated by sunlight – by contracting their pupil into a narrow horizontal slit.

F. D. Hanke, W. Hanke, C. Scholtyssek, G. Dehnhardt **Basic mechanisms in pinniped vision** (2009): Exp. Brain Res.199(3-4): 299-311
F. D. Hanke, R. H. Kröger, U. Siebert, G. Dehnhardt **Multifocal lenses in a monochromat: the harbour seal** (2008): J. Exp. Biol. 211(20): 3315-22
K. T. Brown **A linear area centralis ext. across the turtle retina and stabil. to the horizon by non-visual cues** Vis. Rsrch., Vol. 9, pp. 1053-1062. Pergamon Press 1969

Deceased Galapagos sea lion

Tear fluid without tear ducts

The cornea represents the outer boundary of the eye. It is steadily covered by tear fluid that washes away foreign bodies and prevents direct contact with salt water. Unlike their close cousins, however, these animals do not possess tear ducts.

Turtles and the unusual control over their eyes

The well-developed eyes of turtles provide them with color vision. Even in low-light situations, they are able to see very well, if only at very short distances. Some species exhibit a unique characteristic, being able to rotate their eyes about the axis that connects both pupils (as pictured in the photographs of the cooter below) so that the central line always remains horizontal – presumably, to maximize imaging clarity. Their sense of balance, residing at the center of their brain, controls the muscles that accomplish this feat.

Cooter *(Trachemys scripta scripta)*

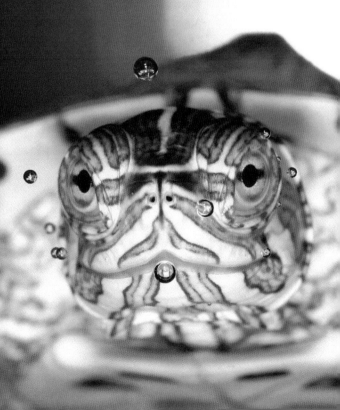

Short-sighted by nature

Moving instead of deforming lenses
In their "default position", most fish can only see distances of up to one meter. In contrast to our own eyes, their spherical, fixed lenses cannot be deformed and thus are unable to adjust their focal length to focus on objects at varying distances. The spherical lens must accomplish this refractive task by itself, as the cornea cannot properly act as a refractive surface due to its refractive

Butterflyfish
(Chelmon rostratus)

Blowfish (Tetraodon schoutedeni)

index being so similar to that of water. This accounts for its relatively large size and for its inhomogeneous refractive indices that afford these animals a spherically well-corrected and luminous image across the whole field of vision. In focusing on more distant objects, fish simply pull the entire lens further into the eye by the use of special muscles. Even assuming optimal underwater conditions, it is only possible to see about 20 m ahead. Thus, sharp vision at long distances is irrelevant for creatures living in water. Land vertebrates – and mammals in particular – exhibit good vision at distances of many kilometers.

Peculiarities of the cornea and retina
After only ten to thirty meters under water, sunlight is reduced to merely one percent of its original energy. Ninety percent of water – whether in lakes or in oceans – is engulfed in perpetual darkness. Despite this dim aspect of aquatic life, fish have a relatively well-developed sense of vision. Some deep-sea fish compensate for the low amount of residual light with especially large eyes. The cornea of fish is often tinted yellow, which changes the bluish-yellowish background brightness in the retina and increases the contrast of bright objects. Many species of fish are even able to pull their retina cells back when swimming through bright waters, using muscle-like fibers in their pigment layer (retinomotor change).

Considerable visual performance
The seemingly simple eyes of fish are capable of remarkable feats of vision. Their advantage lies in their complete panoramic view, depending on the orientation of the eyes. This is necessary, as fish – lacking any morphological feature resembling a neck – are unable to move or rotate their heads to change their field of vision. Like our own eyes, they have rod cells for the detection of brightness and darkness, and cone cells for the detection of colors. The ratio of rod cells to cone cells depends on the depth in which a species dwells.

An increase of photo receptors at great depths
To fit the largest amount of visual cells into the available space, deep-sea fish possess up to twelve layers, one superimposed on the other. Many species further increase the density of receptors by so-called dual cones. Moreover, a tapetum lucidum behind the retina reflects transient photons back to the visual cells for an additional round of stimulus. Such remarkable residual light amplification allows deep-sea animals to see in almost complete darkness, although not very sharply.

Fish at the surface

The problem of short-sightedness

Many fish are short-sighted in water, and would be even more so if they ventured above the surface. Flying fish, wishing to avoid the gaping jaws of their predators, exhibit a deformation at the lower part of their eye which counteracts short-sightedness in the manner of eyeglasses. Fish that prey close to the water surface are also affected by similar survival pressures. The four-eyed fish (*Anableps anableps*, top right) inhabits the surface of brackish waters of South America and needs to be aware of what goes on in the surrounding air and water. The iris of these fish splits their eyes at water level into one half that peeks above the water surface, and another that is submerged below it. Horizontal flaps on the iris prevent the light that scatters from the water surface interfering with above-water vision. Their additional iris gives these eyes a subdivided appearance – hence the term *four-eyed fish*. To eyes of this sort, the world above the water appears shrunk by a pronounced barrel distortion, similar to the image produced by a fish eye lens. All straight lines that do not pass through the image center appear to be curved.

M. A. Ali **Sensory ecology** (1978): Plenum, New York
E. R. Baylor **Vision of Bermuda reef fishes** (1967): Nature 214: 306-307
J. D. Pettigrew, S. P. Collin **Terrestrial optics in an aquatic eye: The sandlance *Limnichthyes fasciatus*** (1970): J.comp.Physiol. 177: 397-408

Mudskipper
(*Periophthalmus spec.*)

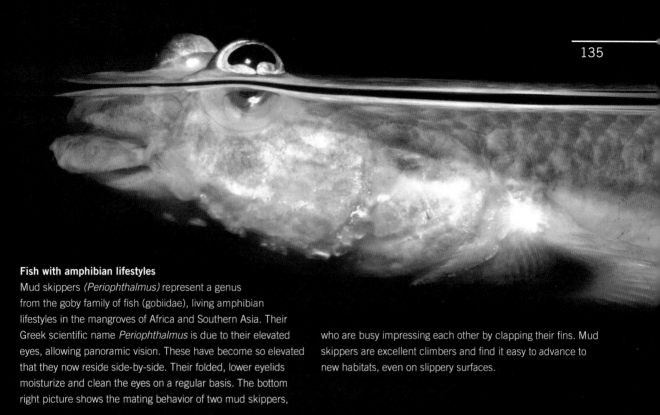

Fish with amphibian lifestyles

Mud skippers *(Periophthalmus)* represent a genus from the goby family of fish (gobiidae), living amphibian lifestyles in the mangroves of Africa and Southern Asia. Their Greek scientific name *Periophthalmus* is due to their elevated eyes, allowing panoramic vision. These have become so elevated that they now reside side-by-side. Their folded, lower eyelids moisturize and clean the eyes on a regular basis. The bottom right picture shows the mating behavior of two mud skippers, who are busy impressing each other by clapping their fins. Mud skippers are excellent climbers and find it easy to advance to new habitats, even on slippery surfaces.

Mudskipper *(Periophthalmus spec.)*

What do whales and hippos have in common?

Hippopotamus *(Hippopotamus amphibius)*

Pilot whale *(Globicephala melas)*

Hippos have large eyes that sit on elevated protrusions at the side of their heads, similar to other aquatic vertebrates that frequently swim on the surface. Like crocodiles, their heads often peek above the water surface while the rest of their bodies are submerged. Their elevated nasal and eye regions allow the animals to breathe and to enjoy a panoramic view. Today, only two species remain in Africa, the hippopotamus and pigmy hippopotamus. The latter is only found in the Niger delta in the south of West Africa, while the large hippo can be found in fragmented parts of Western, Eastern, and Southern Africa. Until quite recently, the larger hippopotamus lived across all of sub-Saharan Africa and could also be found roaming the upper parts of the Nile – right up until the Nile delta.

In recorded history, Madagascar used to be home to up to three species of hippos, which went extinct as the island became populated by humans. Other species of hippopotamus could also be found in Europe and Asia until the late ice age. The pigmy-sized versions of these animals are especially famous, growing to a shoulder height of just under one meter, and inhabiting various Mediterranean islands (like Sicily, Malta, Crete, and Cyprus) before expiring towards the end of the ice age. Contemporary hippopotami share their closest ancestral kinship with whales – a rather counterintuitive conclusion that was reached following molecular and genetic studies.

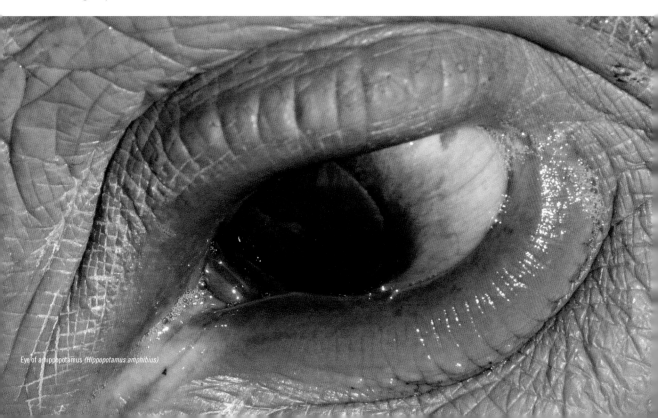

Eye of a hippopotamus *(Hippopotamus amphibius)*

The eyes of aquatic mammals

B. A. E. van der Pol, J. G. F. Worst, P. van Andel **Macro-anatomical Aspects of the Cetacean Eye and Its Imaging System** (1995): In: R. A. Kastelein, J. A. Thomas, P. E. Nachtigall (eds.) **Sensory Systems of Aquatic Mammals** pp. 409-414. Woerden, The Netherlands. De Spil Publishers

Common dolphin *(Delphinus delphis)*

When opening our eyes under water, we witness a blurry world, because our lens system is adapted to vision through air – the only medium in which the refractive system of the eye (cornea, intraocular fluid, lens, and the vitreous body) provides a sharp image. Due to different refractive indices at play, we are extremely far-sighted under water (about 50 diopters). To see the underwater world properly, we must help ourselves with diving googles, so that our eyes are surrounded by air. Due to the planar mask at their front, distances appear shortened by 3/4, while all objects appear at 4/3 of their actual size.

We already know from page 128 that fish have excellent underwater vision – aquatic mammals, however, descend from terrestrial ancestors. Their lenses are more spherical and less elliptical than those of their land-dwelling cousins. This allows light to be much more strongly refracted, compensating for the lack of refraction that occurs through their cornea. As dolphins or seals peek out of the water, they are thus faced with the opposite problem: on land, everything appears blurry to them, and their round lens is better suited to compensate for these distortions. More information on the eyes of seals is available on page 130.

Seal *(Phoca vitulina)*

Beluga whale *(Delphinapterus leucas)*

How large can or should eyes be?

Whale sharks and great whales have eyes roughly the size of tennis balls. Mathematical models that compare the visual acuity of different sizes of eyes at different water depths have shown that larger eyes tend to be more sensitive to luminosity. However, from a pupil size of 2,5 cm onwards, the visual acuity increases only very slowly. Therefore, it appears that for most species, very large eyes are simply not worth the biological effort.

The eyes of the giant octopi, however, are as large as soccer balls – the whole animal being up to five meters in length and equipped with tentacles up to eight meters long.

Bigger eyes are better able to detect large, luminous objects at depths of over 500 m – especially the light paths of sperm whales which prey on octopi, whose luminous displays are produced by bioluminescent algae and aquatic animals.

Whale shark *(Rhincodon typus)*

Hot spots – targeted mutations

By targeted, artificial selection of individual animals that happen to have desirable properties or mutations, such as the especially large eyes of the pictured veiltail – a variation of goldfish – breeders are able to propagate animals far different from their wild ancestors. Such mutations are embedded in the germ cells, and are, thus, heritable.

Telescope eye goldfish,
a breed of regular goldfish
(Carassius gibelio f. auratus)

If the selection procedure is repeated across many generations, strange beings can sometimes emerge. Consider a tiny lap dog, which descends from wolves and to which, in theory if not in practice, it remains reproductively compatible. The pictured lionhead retains fully functional fish eyes with all of their special characteristics, such as a spherical lens that is able to change its distance to the cornea in order to focus on different distances. If the whale shark on page 141 had eyes of comparable relative size, they would be at least one meter in diameter. Their light acuity would, of course, be enormous ...

7 Pax and homology

The third act in the history of evolution

Genes for the development of light-sensitive cells can be found in the first single-celled organisms. From such early forms, the animal kingdom has evolved many types of eyes, all of which make use of common genetic functionality during embryonic development.

Pax and homology
the third act in the history of evolution

Act one: What is evolution?
In 1856, *Charles Darwin* was the first to formulate a consistent theory of evolution by discovering the main causal factors that produce evolutionary changes in organisms, these being selection and mutation. The variety of individuals in a population provides the raw materials for a selection procedure that results in some individuals being more (and others to be less) successful at reproduction, based on momentary environmental conditions. This changes the genetic makeup of the population that follows – the quintessence of evolution.

Act two: Modern evolutionary synthesis
The first half of the 20th century produced the so-called modern evolutionary synthesis – a combination of Darwinian theories with the laws of heredity. The evolutionary mechanisms described by Darwin – mutation, variation, heredity, and natural selection – act primarily on the level of genes and produce new adaptations and even new species over time.

Act three: EvoDevo
An entirely different question concerns the change from an inseminated egg with a certain set of genes to a fully differentiated lifeform with a head, arms, legs, wings, and other body parts – all in their appropriate place. Modern genetics provided great advancements in this field, as did the synthesis of evolutionary theories with those of developmental biology. In fact, such deliberations have opened up an entirely new field of research that is now called evolutionary developmental biology (often abbreviated as EvoDevo).

All organisms are based on identically structured DNA
The number of possible genes may be very large, but not infinitely so. Some genes of microorganisms barely differ from those of humans. EvoDevo concerns the question of how such a stunning array of species could develop from a limited number of building blocks. Researchers have since discovered that the

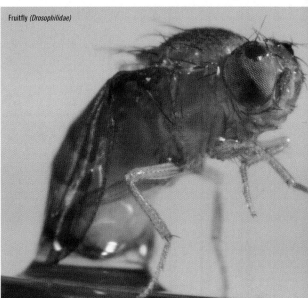
Fruitfly *(Drosophilidae)*

successive activation of genes in the embryo is subject to strict hierarchical control.

The time of activation is important
Due to the different times at which certain genes are activated, it is possible for a single gene to have different effects in different species. So-called master control genes are tasked with providing the impulse for the formation of entire organs.

Hox and Pax genes regulate the fundamental body plans
The so-called Hox genes specify the length axis of the organism. At a later stage, the so-called Pax genes are of fundamental importance for the development of eyes. The Pax6 gene encodes

House mouse
(Mus musculus)

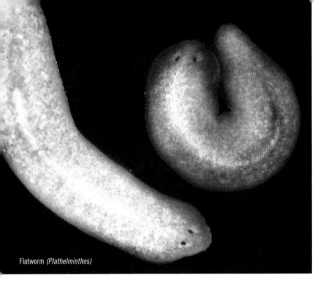
Flatworm *(Plathelminthes)*

a protein that regulates the transcription factor, in other words, the activity of subordinated genes. It was soon discovered that a group of Hox genes can be found in virtually all examined multicellular (metazoan) animals, and that it may, therefore, be classified as homologous. For this reason, these genes must have been conserved over large spans of evolutionary time – at the very least since the Cambrian explosion, dating back 540 million years.

Homeotic genes
These genes control the identity of segments and the detail of their structures. Mutations among these genes can lead to a complete or partial change of organ structures, or even of whole bodily segments. Homeotic genes were discovered due to a very conspicuous mutation in fruit flies (*Drosophila melanogaster*), which grew legs on their heads in places where their antennae should normally be. These mutants were then called "antennapedia".

The discovery of key genes
A homeobox is a characteristic sequence of such homeotic genes. It codes for special, delimited areas of proteins and is able to shut down whole areas of genes. From the genetic analysis of spontaneous mutations that caused fruit flies to develop without eyes, researchers were able to identify the so-called key gene based on a regulatory cascade during the development of eyes. This Pax6 gene turned out to be important for further development, as it strengthened the transcription of some genes, while restricting the transcription of others. It belongs to a whole family of regulatory genes that control a variety of developmental processes.

Functioning eyes in the wrong place
A spectacular experiment succeeded in artificially inducing the development of additional, functioning eyes in other parts of the fruit fly body – for instance, at the antennae, at the wing basis, and on the thorax. As the genomes of many other organisms were compared – a feat that is almost routine these days – this Pax6 gene turned out to be homologous. Even animals with only very primitive eye spots (such as the primordial flatworm *dugesia spec.*, or the box jellyfish that have lens eyes at the side of their shields) were found to be under the influence of the same or of a homologous gene. This is remarkable, as the animals found today have diverged since at least the Cambrian era 540 million years ago. Despite this enormous difference in ancestry, it was possible to produce additional eyes in fruit flies using a version of the gene that originated in mice.

The evolution of complex diversity
Many such control genes – responsible for the development of an organ and compatible with wildly divergent species – have since been discovered. These "main switches", which govern the activity of several hundred subordinated genes, also provide an explanation for one of the most difficult problems of evolutionary biology, that being the emergence of complex diversity.

Even single-celled ancestors have evolved a light-sensitive protein
All of these discoveries led to a new understanding of evolutionary processes in developmental biology, and may therefore, be called the *third act* in our quest to understand the history of evolution. From comparative morphology, one previously had to assume that the diversity of eyes in the animal kingdom must have evolved independently and therefore convergently. The developmental geneticist *Walter Gering* suggested a hypothesis that envisioned the existence of a common genetic basis for all types of eyes – the Pax6 gene. This would indicate a non-convergent evolutionary history of eyes. Gering claimed that light-sensitive proteins had already evolved in single-celled ancestors.

Two types of light-sensing cells
From this ancestral material, a cell that specializes in light detection could evolve in the first multicellular animals. It carried cilium, as well as a membrane in which the photosensitive protein was embedded. The membrane was then enlarged by protrusi-

ons in an effort to increase the space available to that protein, and thus, to maximize light sensitivity. This type of receptor cell is classified as "ciliar". Such types of cells are common in the eyes of most animals, including humans. Arthropods have developed a different type of sensory cell. They have reduced the cilium, enlarging the cell membrane instead through finger-like protrusions, known as microvilli. This produced a light sensory cell that, if aggregated into groups, formed a rhabdom. Rhabdoms function as conductors of light, and can mostly be found in the ommatidia of crabs and insects. This type of receptor cell is classified as "rhabdom-based".

The prototypical eye

Gehring termed an organ a "minimal eye" if it only consisted of one photo receptor cell and another cell to block incoming light. Such a photoreceptor is also called a prototypical eye. Being equipped with the pigment rhopsidin, it is already guided through its development by the Pax6 gene. Darwin himself has raised the question of how often independent eyes evolved in the various groups of the animal kingdom. He already suspected that all eyes could have originated from a prototypical version. In fact, such types of simple eyes were later discovered in flat worms. Gehring, among others, later proved this assertion, by concluding that the development of all types of eyes is controlled by the Pax6 gene – from simple pit eyes, to pinhome camera eyes, lens eyes, and the huge diversity of facet eyes.

Pax6 is in control

Molecular studies have shown that the procedure by which eyes develop is the same in all animals – from flat worm to mammals, including humans. About 65% of genes expressed in the retina of the fruit fly are also active in the retina of the mouse. The Pax6 master control gene stands at the top of the hierarchical cascade. Even the eye's two-cell prototype, which can still be found in some species of flatworms, is controlled by the Pax6 gene. This system of cascading control occurs likewise in all higher lifeforms, leading to the development of very different types of eyes by the introduction of more and more genes – for instance, genes for the development of lenses.

Opsines cause the nerve impulse

Despite this morphological diversity, similar physical and chemical principles determine the process of vision. Opsines are present as visual pigments in all multicellular organisms. These are photoreceptor proteins that when stimulated by a single light impulse (photon), experience a change in conformation, causing a nerve impulse to fire.

Why do some large groups of animals lack eyes?

As elegant as the explanation of the embryonic development of eyes under control of Pax6 genes may seem, it does not address yet another seeming discrepancy. Indeed, many greater groups of animals lack eyes altogether, while others have developed very different tpyes of eyes. Whether all of these types are homologous or not depends on the correct interpretation of homology, including its criteria.

Homology – analogy

Organs, structures, physiological processes, macro molecules, and even behavioral patterns are termed homologous insofar as they exhibit such similarities in two or more species that they must have inherited said features from common ancestors. A statement about homology implies a dedicated assumption about the ancestry of the observed features.

How to recognize homology

Three criteria must be met for homology to be properly asserted:

1. The criterion of position

When comparing specific structures of different species, they should occur in analogous locations. The bones of our limbs are a very simple example of this relationship. In the front limbs of all mammals, the radius always resides between the upper arm bone (humerus) and the wristbone (carpus), regardless of

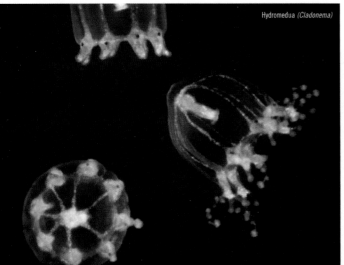

Hydromedua (Cladonema)

whether the limb is dedicated to running, digging, grasping, or for the purpose of flight. Due to a shared genetic basis at play, the construction of these features during embryonic development proceeds in the same manner, and thus, these organs can be considered to be homologous insofar as they are based on the same genetic information inherited from the last common ancestor.

2. The criterion of specific qualities
Different organs may also be considered to be homologous if they are found to be separated from the original group. For instance, an upper arm bone is recognizable as such, even if it is not grouped with the other bones of the limb. A connoisseur is always able to identify the bone as a humerus and can often recognize it as being the humerus of a dog, a cat, or a shrew. This is possible because some organs and structures exhibit a special quality that makes them unmistakable to the experienced eye.

3. The criterion of continuity in alteration
The aforementioned criterion is also true for macro molecules, as it is for genes, whose specific qualities lie in the sequence of their DNA. Since organs and structures may undergo radical change over evolutionary time, a simple comparison between two species does not usually lead to a useful result. Additional species, which exhibit transitional forms, have to be considered as well. Thus, the third criterion is often called the criterion of transitional sequences. It represents the realm of comparative anatomy and physiology, which deals with the gradual changes that occur across the history of species. Usually, it involves the analysis and comparison of contemporary or fossil forms. Such methodology led to the discovery of the phylogenetic history that connects equine animals to contemporary whales.

Not all types of eyes are homologous
The comparison of genes represents a modern method of phylogenetic research, and the conclusions that were drawn from the analysis of Pax7 genes provide a good example of this procedure. However, this does not mean that all eye types within the animal kingdom are homologous. For the development of sophisticated eyes, a cascade of further genes is required, and there is no doubt that these genes evolved much later, as well as completely independently. The Pax6 genes form the basis for the development of eyes, but the question concerning the type of eye that ultimately emerged depends on subsequent adaptations that led to independent evolutionary lineages. This means, that a type of eye is only homologous with respect to other eyes of the same type and complexity. Facet eyes from two species are only homologous insofar as their complexity is comparable.

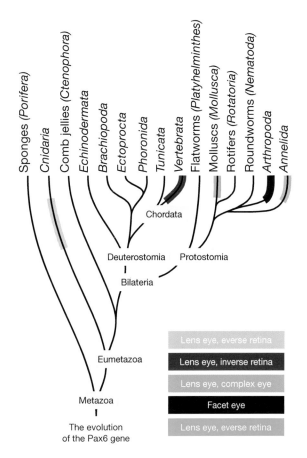

L. v. Salvini-Plawen **Photoreception and the polyphyletic evolution of photoreceptors (with special reference to Mollusca)** (2008): Amer. Malac. Bull. 26: 83-100

8 Alternative senses

When eyes are not enough

Animals inhabit a sensory world that mirrors the challenges that life puts before them. Apart from vision, the senses of smell, taste, sound, feeling, and temperature are ubiquitous among animals. Some are even able to detect electrical and magnetic fields.

Alternative senses

Senses are the arbiters between the exterior world and the neuronal interior world

Dogs experience their world primarily through their noses. Nocturnal bats use ultrasonic echolocation to construct a mental model of their environments. As predominantly visual mammals, we find it difficult to step into the sensory shoes of different animals, so to speak. For their survival, each species uses a specific modality of senses adequate for their environments and lifestyles. In the darkness of caves, eyes are useless, and are, therefore, often supplanted by a sense of touch. Thinkers as early as *Aristotle* identified five humans senses: 1. The visual sense of sight (photoreception). 2. The aural sense of sound (acoustic reception). 3. The sense of smell (olfactory reception). 4. The sense of taste (gustatory reception). 5. The sense of touch (tactile reception). In the remaining animal kingdom, one may encounter many additional senses, such as perception of electrical and magnetic fields. Today, we recognize the perception of temperature as a sense of its own, separate from the sense of touch. The proverbial "sixth sense" of humans may be described in terms of our subconscious deliberations and sensations that are not attributable to any concrete causality.

Mechanoreception

Senses can also be classified by the types of stimulus that they are able to receive and relay. Photoreceptors detect light by absorbing the energy of an incoming light stimulus (a photon) using a visual pigment called rhodopsin. The senses of hearing and touch are related, and are often grouped under the term mechanoreception as they react to mechanical vibrations, pressure or tensile stress. Hearing involves the perception of longitudinal waves that induce oscillation in membranes (eardrums) or in hair and bristles. Ear drums (tympanal organs) – with or without external ears – are the instruments of hearing not only in most terrestrial vertebrates, but also in some insects (grasshoppers, crickets, and cicada). The ears of vertebrates are located on the side of the head. Grasshoppers and cicadas have a pair of tympanal organs at the sides of their torso, while crickets have theirs on the front legs. Spiders and some species of insects are even equipped with hair capable of perceiving sound. These so-called trichobothria are very thin bristles that are brought to resonant motion by sound waves. They can be so sensitive that they are able to detect the faintest turbulences of air. The hairs on the antennae of mosquito males are especially remarkable, as they are able to detect the sounds made by flying females (see page 156). The sense of oscillation (see page 117) is related, as it is able to detect vibrations through solid media (soil, foliage, branches). Insects that run on the water surface (such as the skimmer) measure the deflection of their legs by tiny water waves caused by prey insects falling into water. Specimen from the family of whirligig beetles (gyrinidae) produce such water waves themselves by moving their legs in a circular motion on

Mexican redknee tarantula *(Brachypelma smithi)*

Horseshoe bat *(Rhinolophus spec.)*

the water surface. These waves are then reflected along nearby swimming objects (such as peers, prey, and obstacles). As the reflected waves are picked up by the antennae of the beetle, they are meticulously analyzed by the animal's brain. This process works not unlike the echolocation of bats, who are able to model their environment by an analysis of reflected ultrasonic waves that they themselves produce through clicking sounds. Bats are thus able to "see" with their ears – a skill that can, to some degree, be learned by blind humans.

Chemical senses

Smell and taste are chemical senses, as they involve the direct detection of molecules in gaseous (olfactory reception) or

Maybug *(Melolontha spec.)*

dissolved states (gustatory reception). The detection of humidity (hygroreception) represents another special case. For many animal species, the detection of smell is dominant in their sensory world. This is especially true for mammals, who were originally nocturnal. The olfactory epithelium of mammals lies in their nasal cavity and includes olfactory and supporting sensory cells – bipolar sensory neurons that represent the synapses in the olfactory bulb of the brain. The cilia tips of the bipolar neurons include the olfactory molecules in their membranes and protrude from a layer of mucus, which covers the olfactory epithelium and includes proteins that bind odorant molecules. Many vertebrates possess a special olfactory organ at the roof of their mouths that serves the detection of sexual pheromones emitted by potential mates. Thus, during the so-called activity of flehming, the smells are directly transferred to the olfactory epithelium. Reptiles – lizards and snakes in particular – may only have poorly developed olfactory glands, but are equipped with a special organ – the so-called vomeronasal organ – at the roof of their mouths. They gather olfactory molecules by darting their tongue in and out, thereby leading the molecules towards this organ's sensory epithelium.

The sense of smell of insects

The olfactory perception of insects proceeds by similar principles. Thousands of hair sensilla sit on the insects' antennae and specialize in both mechano as well as olfactory reception. The finely porous hair surface allows olfactory molecules to pass towards the sensory cells.

Communication through scent plays a crucial role in most insects. When looking for partners, females release species-specific sexual pheromones that can be detected by males across large distances. Since males compete for a precise and fast detection of females, their antennae are often strongly enlarged to maximize the number of sensilla on their surfaces.

Special thermal receptors

Some groups of snakes are equipped with organs (containing a thin membrane) lying on both sides of the front upper jaw in an indentation between the nostrils and the eyes. These are infrared receptors able to detect the warmth emanating from the bodies of prey animals – akin to the capabilities of thermal imaging cameras. Pit vipers (such as the American rattle snake

Rattlesnake *(Crotalus spec.)*

and Central American lance-head viper) employ these organs to detect differences in temperature as low as 0.003 degrees. Similar organs have independently evolved along the upper lip of several large snakes (pythons and boas).

Electrical signals

Many species of fish (mainly sharks and rays) are able to detect electrical signals produced by other animals through a very well-developed electroreceptive system consisting of many hundreds of pores, mostly in the proximity of the head. These are the outlets from the ampullae of Lorenzini, which are filled with a gelatinous mass, and whose bottom is covered by electroreceptors that evolved from hair cells. They measure electrical fields by detecting the drop in potential above the skin. The final sensory signal, which may indicate a voltage source, is produced by comparing the neuronal signal of multiple ampullae that cover a single patch of skin. This method of electrical detection only works at close range – at distances below 0.5m.

So-called electrical fish actively produce electrical impulses of this kind and analyze the shape of the thus produced fields in their surroundings. The "electrical images" that these animals are able to see contain information about the shape, distance, and conductivity of the detected objects.

Defensive electrical shocks
However, the aforementioned signals should not be confused with those produced by the electrical organs of certain species that are able to emit electrical shocks of up to 1000 Volts and more than 1 Ampere. Torpedo rays (in tropical oceans), electric catfish (*Malapterurus* in African rivers), and the famous electric

Collared dove *(Streptopelia decaocto)*

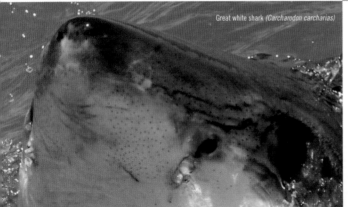

Great white shark *(Carcharodon carcharias)*

eels (*Electrophorus electricus* in the Amazon and Orinoko) are especially notable for their literally shocking abilities, which are used exclusively for defense.

Orientation through Earth's magnetic field
Organs and structures for the detection of Earth's magnetic field have been discovered only relatively recently. Such senses, which are used for spatial orientation, are available to a great

diversity of species such as bees, subterranean mole rats, domesticated pigeons, migratory birds, marine turtles, sharks, and probably also to whales. The most well-known example is found in homing pigeons, who are famously able to return to their nests across large distances. However, the magnetic sense is only one of various techniques that are used for orientation during long migrations. It manifests itself through two separate aspects – one compass-like and the other map-like. The underlying mechanism is only vaguely understood. In some species, neurons sensitive to magnetic fields were discovered with projections

Leopard torpedo *(Torpedo panthera)*

Canadian garter snake *(Thamnophis spec.)*

towards the brain.

Other researchers have concluded that the magnetic sense is localized in the eyes. According to the latter hypothesis, a magnetic receptor consists of a pair of molecules that are activated by light and that form a very short-lived pair of so-called radicals due to a subsequent transfer of an electron. This pair constantly alternates between two possible states of quantum mechanics. The location of this magnetic sense in birds is still not known with great certainty – however, some molecules fit for this task have been identified by researchers, especially the so-called cryptochromes that were found in large numbers (among other places) on the retinas of garden warblers. It is presumed that the cryptochrome might be able to translate a perception of magnetic fields into visual signals, enabling birds to be aware of the Earth's magnetic field.

Another group of researchers has discovered magnetically sensitive structures in the beak area powered by a biogenic magnetite, which counts among the most magnetic minerals. Chains of magnetite crystals arrange themselves with respect to magnetic fields that surround them.

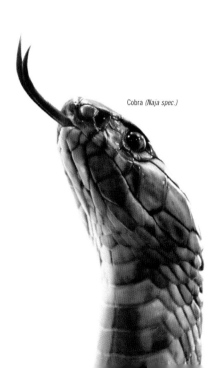

Cobra *(Naja spec.)*

Smell as a replacement for sight

Tiger mosquito *(Stegomya albopicta)*

In the Northern hemisphere, mosquitos represent no more than a nuisance. In tropical regions, however, they count among the most feared carriers of unpleasant and life-threatening diseases – Malaria being merely the most infamous among many. Before laying their eggs in the water, females must ingest a meal of pure blood. They are well-equipped for this task through their mouthparts – perfect instruments for piercing animal skin and extracting a single drop of blood from minute blood capillaries. Their technique for finding an adequate host is as easy as it is effective: a human, for instance, increases the CO_2 content in the surrounding air by exhaling. Being warm-blooded, he or she also leaves a trace of warmth in the vicinity. Both features are picked up by the sensors on the antennae of mosquitos. In daylight, they can use their facet eyes, to detect the shape and location of their victims. At night, all that they need is the change in temperature that alerts them of the presence of a potential host body. Additional hints are picked up via scents (such as the smell of sweat). The precise location of the sting is then decided

Male of the Asian tiger mosquito with its bushy antennae

by tiny temperature sensors on the female's proboscis, since areas of skin directly above blood capillaries tend to be slightly warmer. Finally, a substance to prevent blood coagulation is injected along with the sting, which is also responsible for the itching sensation that we know so well. During the injection of the mosquito's saliva, viruses or single-celled parasites are able to migrate into the bloodstream, explaining the frequent transference of diseases. Male mosquitos are completely harmless, however, as they are only concerned with visiting flower blossoms and drinking nectar. They can be recognized by their strongly feathered antennae that they use to find females.

A male maybug spreads the branches of its antennae by opening its interior blood vessels. This activates the olfactory sensors at the inner surfaces – presumably, to be able to look for food or for mates. When the animal is at rest, it extracts the blood again, which causes the antennae to retract. This protects the delicate olfactory sensors.

Maybug *(Melolontha spec.)*

Sensing touch and vibration

Cucumber green spider *(Araniella cucurbitina)*

Wandering spider *(Cupiennius getazi)*

Lurking in darkness

Spiders have an excellent sense of vibration that is well-adapted to relay biologically important signals. The well-researched spider pictured above *(Cupiennius salei)* belongs to the family of armed spiders *(Ctenidae)*, which do not build nets. Instead, they wait for passing prey in complete darkness. For half a century, they have been considered as model organisms for mechanoreceptive systems, in which sophisticated mechanisms are tasked with preparing the stimulus for the sensory cells. Their pursuit of prey fully depends on the mechanical detection of vibration and air turbulence. If the pattern of stimulus is appropriate, the lurking spider can grab its prey (mostly insects) during jumps through the air.

Hair-assisted hearing

The detection of pressure is accomplished by over 3,000 membrane-covered slits that span the whole surface of the spider's chitincuticula. They are located near the joints of the eight legs and are able to detect forces of Micronewton magnitude by a deformation of the lyre-formed organs' slits. These spiders are able to measure ground displacements as small as 4.5 nanometers – a feat of which contemporary mechanical science can only dream. In addition, each of their eight legs are covered with up to 100 fine hairs able to detect the faintest movement of air. These "hearing hairs" react to the turbulences caused by flying insects and alert the spider to jump and catch its prey at the appropriate moment.

F. G. Barth **Vibrationssinn und vibratorische Umwelt von Spinnen** (1986): Naturwissenschaften 73 (9): 519-530
F. G. Barth **A Spider's World: Senses and Behavior** (2002): Springer-Verlag, Berlin-Heidelberg-New York, p. 394

Bats up close

In contrast to the much larger flying foxes, bats are able to "see" with their ears. Thanks to their sophisticated and subtle ultrasonic echolocation, their cognitive conceptions of space and objects are probably comparable to our own mental models produced by our eyes. For this reason, the eyes of bats are relatively small in relation to their bodies. Within the suborder of bats (microchiroptera), representatives of horseshoe bats emit ultrasonic signals through their nose, while the pipistrels common in Central Europe accomplish the same through their wide-open mouths.

Wikipedia **Pipistrellus** http://en.wikipedia.org/wiki/Pipistrellus

Pipistrel *(Pipistrellus)*

An optimized sense of smell

A split tongue provides olfactory support

At a mass of 70kg and a length of up to 3m, Komodo dragons are the largest of all goannas, and can be recognized by their long tongue that is prominently split at its tip. As in snakes, the tongue can be moved back and forth to aid the most important sense of goannas – their sense of smell. One would be well advised to be careful around these seemingly relaxed animals, as a single bite is enough for the injection of their venom, which rather than being transmitted through hollow or furrowed teeth, mixes with other substances in their oral cavity. Presumably, all goannas are equipped with elongated poison glands below the teeth along each side of their lower jaw.

A successful hunting strategy

The goannas pictured below (on the island of Koh Rok in Thailand) are also two meters long, but not Komodo dragons. Their hunting strategy, however, is quite similar: prey is approached slowly instead of being attacked quickly. This causes prey animals to feel relatively safe. When the venomous bite ultimately occurs, it leads to a slow, excruciating death. Goannas detect their prey by gathering scent on their tongue before drawing the tongue inward, thus propelling the scent particles towards the vomeronasal organ for further analysis. Goannas feel very comfortable in water. The specimen pictured on the right on the island of Rinca near Komodo felt quite at ease inspecting the rubber raft from up close.

Komodo dragon *(Varanus komodoensis)*

Diurnal vipers and nocturnal tree boas

Snakes are primarily visual animals and are thus often equipped with large and light-sensitive eyes. Their nostrils and tongues probe fleeting and non-fleeting substances, respectively. The tips of their tongues are retracted into the mouth towards the vomeronasal organs – two little indentations at the top of their palates – where the scents are analyzed as in the olfactory center of mammals. With both tips of their tongue, snakes are able to detect different scents simultaneously, which provides them with spatial information. This permits them to detect and follow prey animals or potential mates. Thus, the constant movement of their tongues aids them in the detection of their environment. Some species, like the American rattlesnakes, have evolved infrared receptors – indentations between the eyes and the nostrils capable of discerning temperature differences as small as 0.003°C. Similar organs – so-called labial pits located in the rows of scales surrounding the upper and lower lips – evolved in large snakes (boas and pythons). These are not quite as sensitive as the aforementioned pit receptors, but are still able to detect changes in temperature as small as 0.026°C. Both sensory organs are used for the detection of warm-blooded prey, and produce images resembling those of an infrared camera. The green emerald tree boa *(Corallus caninus)* mainly preys on birds that sleep at night and uses its long teeth to hold its feathered victims in a tight grip.

Green mamba *(Dendroaspis viridis)*

T. de Cock **Burning thermal sensitivity as a specialization for prey capture and feeding in snakes** (1983): Am Zool 23: 363-375
A. R. Krochmal, G. S. Bakken, T. J. LaDuc **Heat in evolution's kitchen: evolutionary perspectives on the functions and origin of the facial pit of pitvipers (Viperidae: Crotalinae)** (2004): J. Exp. Biol. 207: 4231–4238.
G. S. Bakken, S. E. Colayori, T. Duong **Analytical methods for the geometric optics of thermal vision illustrated with four species of pitvipers** (2012): J. Exp. Biol. 215: 2621-2629.

Green tree python *(Morelia viridis)*

Two orders of magnitude apart

Many photos show objects divorced from their environmental context, so that it is sometimes difficult to estimate and compare sizes. The pictures on this spread have been taken with the deliberate intent of causing such a confusion. The rose chafer on the left and the young sheep on the right may be quite different creatures indeed, but the photos, nevertheless, draw attention to their similarities. Both animals have eyes at the side of their heads, although their eye construction is fundamentally different. Insects have six legs, while mammals have only four. Do insects have a sense of hearing? Of course, although it works differently from the aural sense of mammals. With the exception of whales, all mammals have outer ears and are able to transmit sonic waves through three

Rose chafer *(Cetonia aurata)*

remodelled bones (malleus, incus, and stapes) before these waves are ultimately interpreted. In both animals, a mouth is used for the ingestion of plant material – rose chafers feed on pollen, while sheep eat grass and herbs. The coat of mammals serves the purpose of heat insulation. The chitin hair of beetles is sensory and also serves the transmission of pollen from one flower to the next.

Young sheep *(Ovis spec.)*

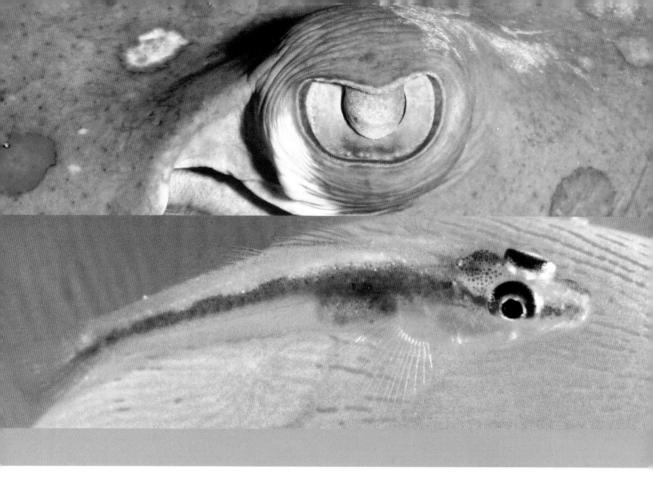

9 A world of color

On the meaning of colors

In all animals capable of color vision, color signals play important roles. Their eyes are adapted to prefer specific hues when selecting mates or searching for food. Depending on their number of receptors, they may even experience colors hidden from us.

A world of color

The sensory perception of color
Color is a sensory perception caused by light of a specific wavelength (or a specific combination of wavelengths) coming into contact with the retina. Light-sensing cells automatically register a stimulus by electromagnetic radiation and photons. The stimulus then travels towards the brain, where it appears in our consciousness as the attribute of color by a process that is still largely not understood. It can, therefore, be argued that color is not a physical property of the observed object at all.

Color refers to reflected light
The term color only makes sense to a living organism possessing light-sensory cells. Thus, the phenomenon of color is not localized on the surface of the object, but rather represents an attribute of its image produced by our brain and presented to our consciousness. We deem objects as being colorful if they absorb certain wavelengths of light while reflecting the rest.

Step 1: Activation of the retina
We can differentiate between several types of light-sensory cells: cones (for color perception during the day) and rods (for black-and-white vision in low-light situations). Each cone contains the pigment rhodopsin in its stack of membranes, which absorbs light from a specific section of the electromagnetic spectrum. This pigment – also called visual purple – consists of a variant of the protein opsin combined with retinal (a derivative of vitamin A). Upon absorption of light quanta or photons by the visual purple, the retinal changes its chemical composition by a quasi-inward folding. This triggers a molecular cascade within the sensory cell, prompting it to activate neurons within the retina.

Step 2: Color perception in the brain
The neurons transmit impulses towards the brain through the optical nerve. We perceive light as being brighter when more light quanta are absorbed by the visual pigments – in other words, when the light is more intense, or when the stimulus of individual cones is greater. However, a singular cone type cannot tell, by which wavelength the stimulus occurred, as visual pigments do not respond with equal strength to different sections of the electromagnetic spectrum. Some sections are absorbed better than others, with each type of visual pigment having its own maximum absorption. Color perception may only occur if several types of cones work together – each having different pigments (rhodopsines) and thus different absorption patterns. Therefore, we perceive colors only if at least two types of cones are reacting to a stimulus. Color acuity increases as the number of cone types grows larger.

The evolution of opsins
The different types of cones differ in the protein component of their visual pigment – the opsin. Researchers have been able to construct hereditary trees of the different opsins found in vertebrates by the analysis of genetic sequences. From the perspective of phylogeny, opsins are very old proteins indeed – much older than the currently predominant groups of vertebrates that have evolved some 400 million years ago. Four different genetic lineages of cone pigments have emerged that focus their absorptive capacity on long-wavelength, normal (visible), short-wavelength, and ultraviolet light, respectively.

The evolution of color vision in birds and mammals
Representatives of all large vertebrate groups are equipped with rods on the retina that are optimized for low-light vision. Their photo pigment is also rhodopsin, and resembles a cone pigment that absorbs medium wavelengths – both in its structure and in its absorptive properties.

Birds have four different types of cone pigments in their retina (in addition to their rods). These types of visual pigments already appear at the beginning of vertebrate evolution, which occurred 350-400 million years ago. Phylogenic analysis suggests that before the split into the great classes of vertebrates took place, different types of cones had already evolved, with rods emerging only at a later time. The majority of cartilaginous fish (sharks, rays), reptiles, and birds already have rods as well as four spectral classes of cones at their disposal, which give these animals tetrachromatic color vision.

Reduction of opsine types among mammals
The majority of mammals only have two types of cones and visual pigments (dichromatic color vision), but since they all descended from ancestors with four types, a reduction must have happened at some point. It is reasoned that the reduction occurred due to the nocturnal lifestyle of the first mammals, who avoided daytime when the landscape was still dominated by diurnal dinosaurs. These ancestral mammals may only have had small eyes – their rods, however, were remarkably sensitive. Two types of cones, which only allowed limited color vision, were

sufficient for their occasional daytime excursions. It was only after the extinction of the dinosaurs 65 million years ago that the ancient mammals were split into groups of very different lifestyles. Many mammals (including most types of cats and dogs) retain two types of visual pigments. Some (like whales and rodents) retain only one pigment, and are thus incapable of discerning colors. The color perception of simian mammals differ substantially from that of most other mammals. Old-world apes, including humans, have trichromatic vision – the detection of ripeness levels of fruit would be much more difficult without it. Since all apes are social animals, their visual sense plays an important part in social interaction, such as in countenance, or in the finding of mates.

Sea goldi male *(Pseudanthias squamipinnis)* with a common cleaner wrasse *(Labroides dimidiatus)*

Stripes across the eyes

Red is the best camouflage color under water, being absorbed at depths of about 5-10 meters. Aquatic creatures of this color living further than 5-8 m below the surface appear cloaked at distances exceeding several meters. From this point of view, the eye of the *Symphorichthys spilurus* on the left page would be its first feature to disappear from view. Moreover, the contours of its body are complicated by its prominent color pattern, which makes it more difficult for visual predators to detect the animal.

The color patterns that span the body (and run across the eyes) of the beautiful broad-barred firefish (*Pterois antennata*) on the right page appear so vibrant only because they have been illuminated by a flashlight at close-range (below 50 cm). At a depth of only a few meters, the whole fish appears remarkably understated. The blue spots on its enlarged pectoral fins remain conspicuous even at greater depths, and thus draw attention away from the body. The animal uses its pectoral fins to trap tiny prey fish before sucking them into its mouth cavity.

Sailfin snapper *(Symphorichthys spilurus)*

Broad-barred firefish *(Pterois antennata)*

Quirky and venomous

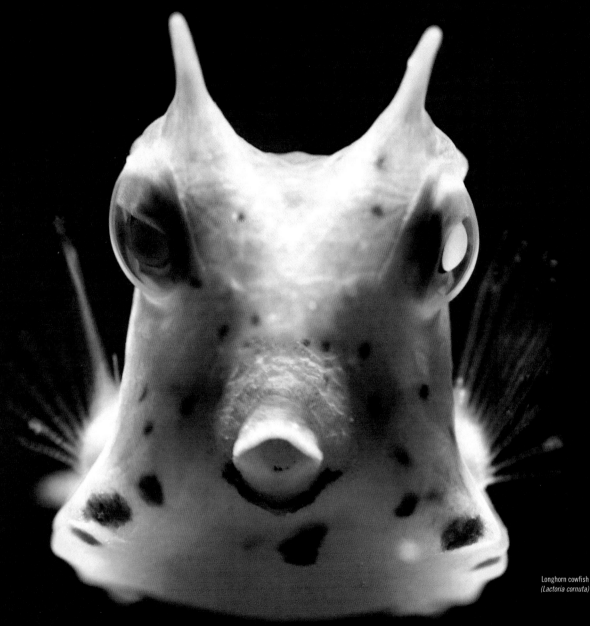

Longhorn cowfish
(*Lactoria cornuta*)

The left photo shows a longhorn cowfish (*Lactoria cornuta*) which belongs to the group of boxfish without a gill cover. They breathe by moving the floor of their mouths up and down. Their skin is highly venomous, making them indigestible to predators – a fact that is advertised by their stinger-like horns, which might become locked in the predator's mouth. The red lionfish on the right page hunts in twilight, trapping its prey in a corner and pulling it into its mouth. Its back stingers are used for defense.

Red lionfish
(Pterois volitans)

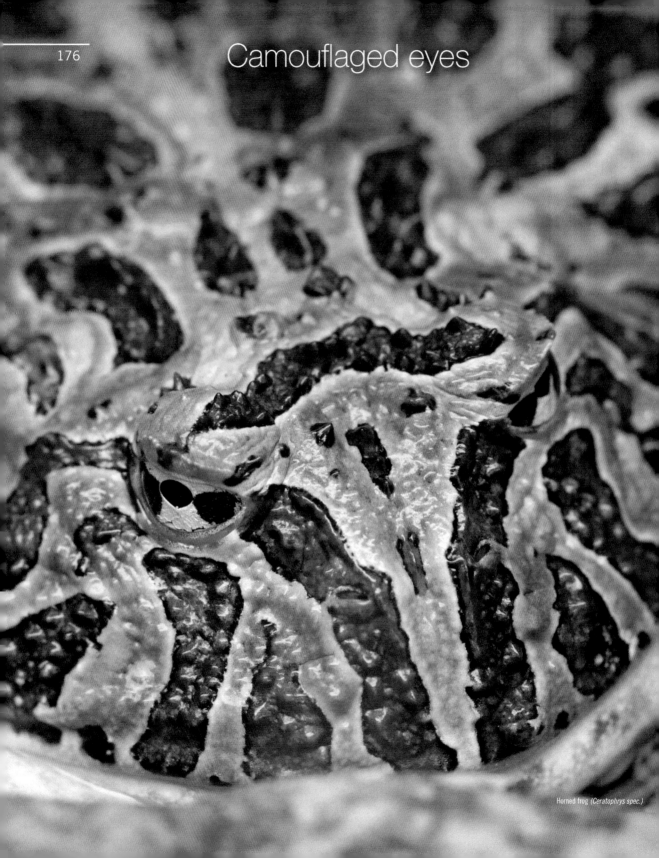

Camouflaged eyes

Horned frog *(Ceratophrys spec.)*

Crocodilefish *(Platycephalus spec.)*

Many vertebrates are visual animals, using their eyes to detect each other's presence, as well as the presence of prey. It is not uncommon for such creatures to classify their predators according to their eyes. Thus, a properly camouflaged animal must also expend some effort in hiding its visual organs. When the sandperch sitting on a coral reef in the left photo hides in the mud looking for prey, only its eyes stick out from the ground (see also page 201). The camouflage in the bottom photo aims at dissolving the contours of the pictured scorpionfish. Red is an excellent camouflage color, being fully absorbed at water depths exceeding 10m. Far from the vibrant display of red shown in the photo below, these fish are completely color-neutral when not illuminated by the camera's flashlight.

The left photo shows a South American horned frog with a pair of particularly well-camouflaged eyes. It lies dormant in foliage or moss, waiting for a prey animal to pass. The victim is then quickly captured through a jump, or by its wide mouth. This species has evolved two dots on each eye (so-called pseudo-irises) to prevent the iris from sticking out from the surrounding striped pattern.

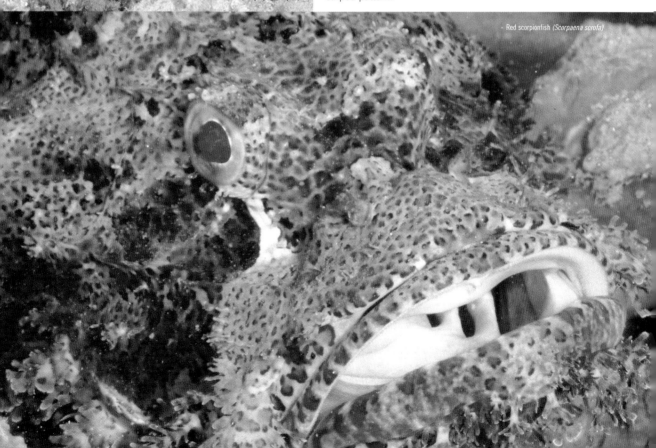

Red scorpionfish *(Scorpaena scrofa)*

The patterns on facet eyes

Ecuadorean tropical fly

H. L. Leertouwer, D. G. Stavenga **Spectral characteristics and regionalization of the eyes of *Diptera*, especially Tabanidae** (2000): Proceedings of the Section Experimental and Applied Entomology of the Netherlands Entomological Society 11: 61-66
O. Burakova, G. Mazokhin-Porshnyakov **Electron microscopy of the compound eye in *Haematopota pluvialis* L. (Diptera: Tabanidae)** (1982): Ent Rev 61: 26–33
G. D. Bernard, W. H. Miller **Interference filters in the corneas of *Diptera*** (1968): Invest. Ophthalmol. Vis. Sci. 7 (4): 416-434
K. Lunau, H. Knüttel **Vision Through Colored Eyes** (1995): Naturwissenschaften 82: 432-434
D. G. Stavenga **Colour in the eyes of insects** (2002): J.comp.Physiol. 188: 337-348

Color pigments or iridescent hues?

Horsefly *(Tabanus bovinus)*

Interference filters

The vibrant colors in the eyes of many gadflies are produced by refractive phenomena – in particular, by a sequence of parallel layers in the cuticula lenses of each individual facet of the complex eye. These function as interference filters, allowing only a specific section of visible light to pass into the eye. The reflected color is the result of interferences that cancel out all other colors. From this, we can conclude that the metallic gold-green eyes of gadflies or lacewings reduce the transference of orange-green hues, thus permitting only certain wavelengths of filtered light to pass towards the onmatidia. These filters are often arranged into stripes that run across the eyes and indicate the spectral sensitivity of the light-sensitive cells below.

Patterns caused by sexual selection?

A diversity of patterns and colors can be found in the facet eyes of the different species of gadflies – including wholly green, striped, and even meandering arrangements. The cause behind these patterns is not well understood, with some theories advocating special visual properties, and others postulating a communication between the sexes during courtship displays. The latter hypothesis explains the color patterns in terms of sexual selection. In fact, there exists a difference among the genders of gadflies. For example, the color pattern of females spans their entire eyes, while the equivalent pattern of males is only found at the lower portion of their eyes.

Common Horse Fly *(Haematopota pluvialis)*

Green lacewing *(Chrysopidae)*

Pseudo-pupils

When taking flashlight pictures of many insect eyes (and of butterfly eyes in particular) we can often see darker regions that do not seem to have a fixed position in the facet eye. Instead, the location of these so-called "pseudo-pupils" changes with the angle of view. Incident light is almost completely absorbed in ommatidia that are nearly vertical with respect to the observer, causing this region to appear black or dark. The "main pseudo-pupil", as shown in the eye of the orange-tip butterfly on the left page, is usually located at the point where light incides vertically.

The remaining pseudo-pupils tend to be distributed in regular (often hexagonal) patterns. The right page shows a Julia butterfly (*Dryas iulia*) during three phases of its departure, with prominent rings of pseudo-pupils surrounding the main spot at the center.

Cabbage White
(Pieris sp.)

D. G. Stavenga **Pseudopupils of compound eyes** (1979): in Autrum, H.: Handbook of sensory physiology VII/6A. pp. 357-439. Springer, Berlin

Julia butterfly (*Dryas iulia*)

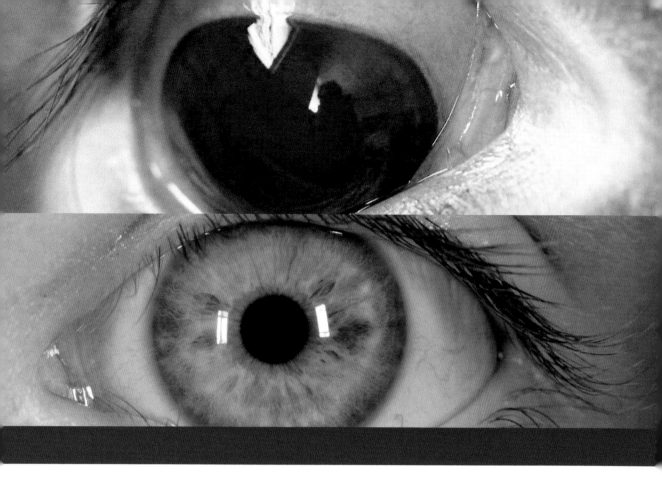

10 The language of our eyes

Of hidden meanings behind our look

To some extent, we understand the eye language of our closer animal cousins. This allows us to communicate many simple emotions across different species. The great apes, being our closest cousins on the tree of life, are able to recognize more of these than other animals.

The language of our eyes

Sibirian tiger *(Panthera tigris altaica)*

More than mere visual organs

Eyes are primarily visual organs for capturing photons and converting them into electrical signals that are then transferred to the brain. However, eyes also have the capacity of transmitting signals that induce emotional states in higher animals like mammals and birds. For instance, the cuteness of children or animals causes humans to spontaneously engage their attention and in protective measures (see page 199). Eyes are especially important in the transmission of such signals. The question of oculesics (the language of eyes) is somewhat more complicated. According to physiognomy, features of our body – in particular, our face, posture, gestures, countenance, and eyes – reflect our emotional state. An insight into human nature largely depends upon the understanding of this subtle language.

The color of the iris

There is a remarkable variability in iris colors among Europeans. Since the iris itself is largely unpigmented, blue is the base tone in most cases. The coloring is caused by a thin layer of pigment on the back of the iris, and shows itself most strongly in babies of a light skin color, as the pigment melanin that is responsible for different eye colors is lacking in young infants. Babies from Africa or Asia are usually born with dark-brown eyes. Young house cats are also born with light-blue eyes and attain their final eye color only after three months. Genetic research suggests that all humans originally had brown eyes, and that blue eyes have appeared only several thousands of years ago as results of a mutation. In general, a lack of pigments in the skin of humans is advantageous in regions of only moderate sunlight, since it assists the body in producing the essential vitamin D.

Crying

Humans are the only mammals to lose more tear fluid than usual during crying. Usually, tear fluid serves to keep the eyes moist, to facilitate the movement of eyelids, and to wash away disruptive particles. The secretion of tear fluid increases during crying by a factor of 400. It is also believed that African Elephants shed tears out of joy or grief. However, the proverbial crocodile tears are definitely caused by something else (see page 201).

Ring-tailed Lemus (*Lemur catta*)

False eyes can be life-saving

Many butterfly species, like the South American morpho pictured here, exhibit circular patterns on their wings that are strikingly similar to eyes. Among the group of satyrines, these false eyes appear in varying numbers, mostly on the bottom sides of the wings. Their purpose is to draw the attention of attacking predators (birds and lizards in particular) away from the insect's body. The false eyes are usually found towards the outer edges of the wings, so that, in case of an attack, only the least essential parts are damaged. In fact, it can be observed that injured butterflies are most often damaged at the back edges of their wings. The specimen on the right page must have been attacked by a bird aiming at the posterior

M. Stevens **The role of eyespots as anti-predator mechanisms, principally demonstrated in the Lepidoptera** (2005): Biol. Rev. 80 (4): 573–588

Blue morpho *(Morpho peleides)*

false eye, but has nevertheless survived the encounter due to its deceptive camouflage.

Satyrine *(Nymphalidae)*

Gossamer-winged butterfly *(Chilades pandava)* with a „false head"

False eyes under water

Angler *(Antennarius sp.)*

Blue Spotted Puffer
(Canthigaster solandri)

One fleeting second to escape

Patterns resembling false eyes also occur in a number of fish species, although for varying reasons. Large false eyes are usually intended to give the impression of a much larger predator. The orange-striped Watchman Goby and the pictured Blue Spotted Puffer accomplish this feat by a single large false eye on their back fin.

Mimicry for defense and deterrence

In addition to the aforementioned method of Batesian mimicry, these color patterns may also imitate warning colors flashed by defensive or poisonous species.

Camouflage as aggressive mimicry

Camouflage is a protective coloration that aims to make an animal undistinguishable from its surroundings. Such tactics may be employed for protection against predators, or as measures of so-called aggressive mimicry by the predators themselves to stay invisible to their potential prey. The angler fish on the left, for instance, needs only to open its mouth to suck an unfortunate victim into its oral cavity.

Orange striaped watchman goby
(Amblyeleotris randalli)

Intuitive oculesics

Can you guess what both mammals on this spread were thinking as these photos were taken? How much are we able to discern from their eyes? The Japanese macaque was merely threatening the photographer, who must have moved too close for comfort. Its moderately impressive jaws were retracted mere seconds later as the camera moved farther away. The lion male on the right was provoked by direct eye contact. It certainly meant business, as it started attacking the car's windshield, which fortunately remained steadfast. Contrary to domestic cats, lions have circular pupils (see page 88) in the manner of most other big cats.

Japanese macaque *(Macaca fuscata)*

Lion *(Panthera leo)*

A comet's impact 65 million years ago ...

65 million years ago, as the Cretaceous period gave way to the Tertiary period, a massive extinction event caused half of the Earth's species to vanish from the world. Among dinosaurs, only birds were spared this ultimate fate. It is presumed that this global extinction was caused by the impact of a meteor on the Yucatán peninsula, producing a crater measuring 180 kilometers in width and causing the violent eruption of subterranean lava masses in the Deccan Traps of the Indian subcontinent.

A unique chance for mammals
As the dust subsided, however, ancestral mammals (shrew-like creatures) found themselves in a world without their main diurnal competitors – the dinosaurs. These tiny animals were originally adapted to a nocturnal lifestyle, having evolved a fur coat and a constant body temperature. The lineage from which apes and humans later evolved started with shrews (genus *Tupaia*). These mainly lived on the ground but were already adept at jumping among the branchwood of trees. Their eyes still lay on the side of their heads, making directed jumps and the grasping of branches truly challenging. It was not until their eye positions shifted forwards, enabling front-facing binocular vision that the further evolution of primates – including the great apes – became possible.

Tree shrew *(Tupaia spec.)*

L. Alvarez, W. Alvarez, F. Asaro, H. V. Michel **Extraterrestrial Cause for the Cretaceous-Tertiary Extinction** (1980): Science 208: 1095-1108
D. Archibald, D. Fastovsky **Dinosaur extinction** (2004): In: Weishampel, Dodson, Osmólska (Hrsg.): The Dinosauria. University of California Press.
Wikipedia **Paleogene extinction event** http://en.wikipedia.org/wiki/Cretaceous%E2%80%93Paleogene_extinction_event

Borneo orang-utan
(Pongo pygmaeus)

Close genes and close emotions

Gelada *(Theropithecus gelada)*

We often find our emotions triggered by the human-like way in which monkeys play with their children or young (the pictures show a gelada pavian mother). The whole process resembles human-play with respect to non-verbal communication that can be scientifically ascertained – from body language to countenance. We find it easy to understand these images because these forms of expressions, and the meanings behind them, stem from common ancestors in which these behavioral patterns originated.

Eye language of our closest cousins

The expressive capacities of our eyes aid us in nonverbal communication, making it possible to infer our frames of mind, mood, and even more specific characteristics of our psyche. Eye contact, therefore, represents an important bridge through which nonverbal communication becomes possible. Weak eye contact reduces the degree to which such signals are effective, making it an indispensable means in the expression of body language (countenance in particular). For this reason, eyes are often considered to be the mirror of the soul. We can even use some of these capabilities to communicate with animals – for instance, direct and wide-open eye contact counts as an aggressive and threatening gesture, even among humans. This almost certainly represents an atavistic evolutionary legacy, as similar threatening gestures have been observed among apes. It is no accident that these animal gestures are intuitively well-understood by us. In *The Expression of the Emotions in Man and Animals* written in 1872, Charles Darwin already wrestled with similar questions, trying to classify our correlation of facial movements (countenance) to our emotions as products of genetic rather than of cultural (learned) similarity. He further noticed many parallels between the emotional expressions of humans and animals, thus strengthening his theory about the common ancestry between humans and animals.

Borneo orang-utan
(Pongo pygmaeus)

Dog (Canis lupus familiaris)

C. Darwin **The expression of the emotions in man and animals** (1872): London, John Murray

Cuteness

K. Lorenz **Die angeborenen Formen möglicher Erfahrung** (1943): Z. Tierpsycholog. 5: 235-409

French bulldog *(Canis lupus familiaris)*

"Cuteness" broadly refers to morphological characteristics of a small child's body – especially of its head and face, such as a large head (and forehead), and the associated low placement of facial features: large, round eyes, a small nose, a small chin, round cheeks, and soft, elastic skin. A child's head tends to be larger in relation to the body, while its limbs are relatively short compared to those of an adult. *Konrad Lorenz* was the first to surmise that evolutionary biology lies at the heart of this phenomenon, whereby cuteness is an adaptation that benefits young children, being the key stimulus that triggers innate responses such as attention, care, assistance, protection, and a decrease in aggressiveness (this occurs most strongly in parents). The morphological features associated with cuteness are so effective that they have been used in many unrelated human endeavors, from advertising to selective breeding. The French bulldog on the right page, for instance, was bred to emphasize a child-like appearance. Even our language betrays the strength of these associations. In German, for instance, species that carry the diminutive suffix "-chen" tend to have short heads with large, round eyes, such as the robin (Rotkehlchen), the squirrel (Eichhörnchen), the rabbit (Kaninchen), the small buttonquail (Laufhühnchen), and the bluethroat (Blaukehlchen).

Ambush predators

Nile crocodile *(Crocodylus niloticus)*

Despite their laid-back lifestyle, Nile crocodiles can very swiftly jump out of the water towards a prey animal.

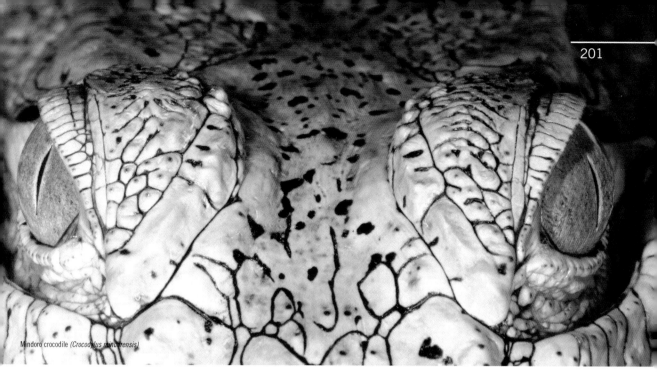

Mindoro crocodile *(Crocodylus mindorensis)*

Crocodile tears

When feeding, crocodiles tend to open their jaws far and wide – so far and wide, in fact, that they experience a momentary shortness of breath, which causes tear fluid to secrete from their eyes. From this observation, it used to be said that crocodiles show mercy for their victims by crying the proverbial tears.

A strange similarity

As an ambush predator, the crocodilefish pictured below (see also page 177) resembles an actual crocodile in its flat shape and its protruding, camouflaged eyes for panoramic vision. The similarly protruding eyes of crocodiles peek above the water surface as the predator's body waits below. The colors of both animals are specialized for camouflage. Despite these similarities, their strategies for dealing with prey animals could not be more different. While the crocodile lunges out, grasping its prey with its enormous jaw equipped with razor-sharp teeth, the fish must merely open its mouth to pull its victim into the oral cavity.

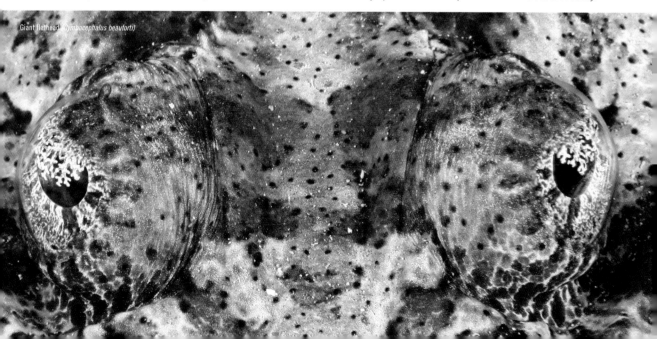

Giant flathead *(Cymbacephalus beauforti)*

The hypnotizing eyes of vipers

Horned viper *(Vipera ammodytes)*

Vipers or adders (Viperidae) are a family of viviparous poisonous snakes living in America, Africa, and Eurasia. In Europe, the species of crossed viper is the most widely distributed. Its native terrain leads up to the Southern Alps, whereupon the asp viper makes its appearance. The Baskian viper inhabits the north of the Iberian peninsula. Spain and Portugal represent the territories of the snub-nosed adder. The horned viper inhabits regions spanning Austria, Switzerland, and South-Eastern Europe (right until the black sea). The field adder is Europe's smallest and rarest. The meadow adder, the Caucasus viper, and the Dinnik's viper inhabit the Caucasus region. Some snakes look rather stocky, having a thick body with a short tail, such as the Gaboon viper *(Bitis gabonica)*. In other species, as evident in the green grass snake *(Opheodrys aestivus)*, the transition between thicknesses is more gradual. Their cross-section also varies, from round to oval to triangular. Snakes do not have eyelids. Instead, their eyes are completely covered by a transparent scale. This is different in glass lizards, which can be recognized from the way in which these animals blink. As evident in the photos, their pupils are shaped like elongated slits in the manner of cat pupils.

Gaboon viper *(Bitis gabonica)*

Cuteness under water

As the pictured blowfish shows, our perception of cuteness also applies to aquatic animals. The large, widely spaced eyes of these porcupine fish may produce associations of "parental protection" in some human observers, despite the actual, highly venomous nature of these animals. Their gonads, liver, and skin (excluding their muscles) contain an infamous neurotoxin that they first consume as part of their diet before depositing it in the aforementioned organs.

Despite the dangers that are involved, these fish are considered a delicacy in Japan, requiring the skilled labor of specifically trained chefs. Today, they are bred without the neurotoxin, making them completely safe to eat. Porcupine fish inflate when sensing danger, making them much harder to swallow.

Blowfish *(Tetraodon biocellatus)*

Black-blotched porcupinefish *(Diodon liturosus)*

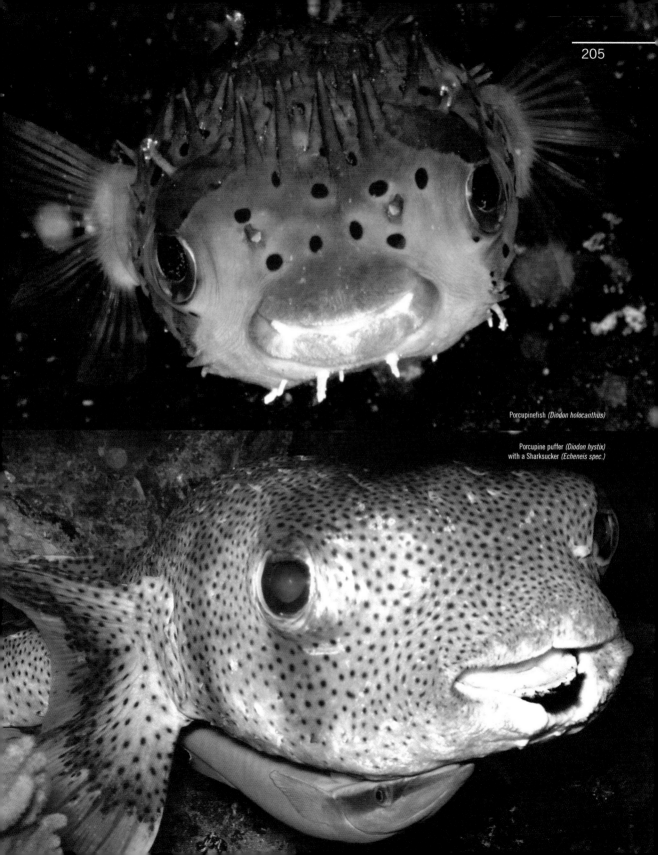

Porcupinefish *(Diodon holocanthus)*

Porcupine puffer *(Diodon hystix)* with a Sharksucker *(Echeneis spec.)*

Teddy bears or beasts of prey?

It is an amusing aspect of our culture that bears, who are justly famous as potentially dangerous predators, are found as fluffy miniatures in many nurseries. When standing upright, they bear a striking resemblance to humans (pun intended) and thus occupy a prominent place in ancient mythology. They have long been considered to be good-natured, portly, friendly, and even emblematic of comfort and motherly care. Even though young bears do not resemble human children very strongly, their thick fur seems to be a source of great attraction for the young. However, the countenance of bears is unrecognizable to the layman, being very different from human expressions. Thus, we are at risk of misunderstanding the aggressive warning signs of these animals.

Brown bear *(Ursus arctos)*

How many species are there?

A lot of vertebrates ...

Around 60,000 species of vertebrates have been identified worldwide. They can be subdivided into six rough categories: mammals (8%), birds (16%), reptiles 12%), amphibians (5%), fish (55%), and primordial vertebrates (4%). This means that only around every twelfth vertebrate species is mammalian. The vast majority of vertebrates are equipped with rather similar eyes – powerful lens eyes, to be precise.

... and a lot more invertebrates

These vertebrate figures are dwarfed by a large and virtually inestimable diversity of invertebrates. Insects alone are split into a million species. Furthermore, around 100,000 species of spiders, 85,000 species of molluscs, and 47,000 species of crabs are assumed to date. They all have either facet eyes or small lens eyes (excepting octopi).

What about Central Europe?

For comparison: only about 700 species of vertebrates have been found in Central Europe, as well as 40,000 species of insects, and more than 5,000 species of invertebrates (excluding more than 2,000 separate species of arachnids). To put it in another way, only about 1.5% of all animal species in Central Europe are vertebrates. A far greater diversity of species

Hummingbird from Ecuador (*Trochilidae*)

Around 330 species of hummingbird are at home on the American continents. They count among the smallest warm-blooded animals.

The giant squid is a remarkable exception among animals, both with respect to its total size, as well as to the size of its lens eyes, which are the size of soccer balls!

exists among insects, which account for 80% of all species. The same rough proportion applies worldwide.

An overview of a wide spectrum

By proportion, vertebrates are more present in this book than in the actual biosphere. Vertebrates are usually larger, making them easier subjects of photographs. They also represent our immediate biological cousins – accounting for their prominent status on these pages. Despite this anthropocentric focus, it is hard to deny the incredible variety of invertebrate animals, especially with respect to their eyes. The eye has convergently evolved dozens of times, producing radically different, but consistently well-adapted results. The Pax6 protein function (see page 146) forms the basis of all eyes and thus occurs in the gene sequences of all animals.

> Among mammals, the red fox is of perfectly average size. Both on land and in water, the class of mammals is by far the heaviest. Without counting humans and ants, mammals contribute the most mass to the animal biosphere.

M. Gleich, D. Maxeiner, M. Mirsch **Life Counts: Cataloging Life on Earth** Atlantic Monthly Press, 2002
G. Giller **Are We Any Closer to Knowing How Many Species There Are on Earth?**
www.scientificamerican.com/article/are-we-any-closer-to-knowing-how-many-species-there-are-on-earth

Red fox
(Vulpes vulpes)

Glossary

Accommodation: A direct change in the refractive strength of the eye lens in order to project a sharp image onto the retina. In related analysis, the near point specifies the shortest, and the far point the longest distance at which sharp imaging is possible.

Adaptation of the eye: Light-sensory cells adjust their intensity depending on the strength of illumination. Adaptations to brightness usually proceed very quickly, while adaptations to darkness take considerably longer. The chromatic adaptation in the retina resembles a "white balancing", or an adaptation to changing spectral situations. In many animals, the adaptation to brightness and darkness occurs by a movement of receptors in the retina.

Apposition eyes: Facets whose ommatidia are fully separated (and isolated) by pigment cells. Such eyes require a greater amount of light, but are able to see extremely sharply.

Auxiliary eye: See main eye.

Blind spot: The polarity of vertebrate retina cells points inwards, because the whole retina evolved from the cerebral epithelum, and because the retina cells were produced by a folding of the epidermis during early embryonic development. The later evolved receptor cilia are inwardly directed, producing an inverted vertebrate retina that points away from light. Axons depart from all retinular cells, and are routed towards the brain at a particular spot – the blind spot – that lacks retinular cells for the detection of photons. Thus, the blind spot is the location on the retina where the nerves proceed towards the brain. Animals with an everse retina – the majority of visual animals (octopuses, arthropods, etc.) – do not have a blind spot.

Bubble eye: A primordial type of eye with an indented retina covered by a transparent epithelium. The indentation is usually filled with a liquid, producing a bubble-like lens.

Chiasma opticum: The intersection point between the optic nerves of the left and the right eye. In apes (and thus in humans), roughly half of the nerve fibers from one side are routed to the other side, where they are connected to the remaining fibers of the opposite side.

Ciliary muscle: During accommodation (dynamic change of the refractive strength of a lens), the lens is brought closer to the retina by a contraction of the corneal muscle (as in primordial fish). In cartilaginous fish and bony fish, the fixed lens is pulled back. In amphibians, it is pulled forwards.

Complex eye: See facet eye.

Cone: A type of retina cell dedicated to diurnal vision in vertebrates. Light-sensitive cones may possess different absorption maxima. Three types of cones exist in the eyes of humans: One for red, one for green, and one for blue vision. With three types of cones, color vision becomes possible.

Cornea, cornea lens: The transparent, curved anterior part of the outer eye that is moisturized by tear fluid in mammals. It represents the outer boundary of the eye and usually partakes in light refraction. The cuticular lens in the eyes of arthropods, also called cornea, is composed of transparent chitin, and is separated from the single-layer epidermis like the remaining cuticula of the entire body. The process of molting involves all corneas of facet eyes. The transparent epidermis above the retina in the eyes of animals, like snails or box jellyfish, represents yet another cornea.

Crystal cone: Part of the dioptric apparatus in the facet eyes of crabs and insects, which are built by four so-called crystalline building cells (Semper cells). It serves as a light-focusing apparatus after the cornea. In the case of apposition or superposition eyes, it focuses light onto the proximal tip, and from there towards the rhabdom. In the case of the mirror reflection eyes of higher crabs, it focuses light along its square inner surfaces towards the rhabdom.

Cuticula: A clipping of the epidermis as an outer protective layer and an exoskeleton of the body. In arthropods, this cuticula is made from chitin – a colorless, non-water-soluble polysaccharide composed of a long chain of sugar subunits (N-acetylglucosamine).

Dichromatic vision: Light-sensing cells may be sensitive to different wavelengths, which are usually distributed among different types of photoreceptor cells. Animals with dichromatic vision possess only two types of such cells (cones) in their retina – most mammals, excepting old-world apes, possess such vision.

Dioptric apparatus: A refractive system in the lens eye of vertebrates and cephalopods, as well as in the complex eyes or arthropods.

Evolution: The change in organisms over time caused by selection.

Facet eye: A complex eye consisting of multiple aggregated eyes (sometimes in very large numbers). The separate eyes are called ommatidia.

Fovea: The fovea centralis is an indentation at the center of the so-called yellow spot (macula lutea) where the concentration of receptor cones is especially large and where vision is sharpest – in humans, this spot measures roughly 1.5mm in diameter.

Frontal ocelli: Point eyes of insects, lying at the center of their heads between the facet eyes. They are equivalent to the median eyes of arachnids, and to the nauplius eyes of crabs.

Larval eye: Many species are equipped with different eyes during larval and adult stages. At first, the nauplius larvae of marine crabs have only median eyes, called nauplius eyes. Larvae of holometabolic insects (such as butterflies, beetles, and flies) have strongly modified remainders of facet eyes that are grown anew after pupation. These so-called stemmata represent an adaptation of visual organs to the larval lifestyle.

Lens: A lens is a transparent optical building block with two refractive surfaces, of which at least one is curved in either a convex or a concave way. Lenses in the animal kingdom include the corneal lenses of arthropods, or the (more or less) spherical lenses in front of the retina (solitary or additional) that form the dioptric apparatus.

Lens eye: A light-sensitive organ composed of either a lone cornea, or a cornea and a lens. The eyes of all vertebrates and most invertebrates belong to this group, excepting arthropods, which primarily have facet eyes.

Lobus opticus, optical nerve: A bundle of nerves originating at the axons of the retina cells and proceeding towards the brain. In higher vertebrates, these axons descend from nerve cells of ganglian cell layers in the retina, which already carry out an initial processing of sensory information.

Main eyes: In real spiders (araneae), the main eyes are descended from the median eyes still present in other arachnids. In many groups of spiders (such as in jumping spiders), they are the most important (and especially large) visual organs.

Median eye: A general term for simple photoreceptors or lens eyes in the central head region of arthropods. Originally, these animals had four median eyes, which were later reduced to three in crabs and insects. Arachnids also had four median eyes at early stages of their evolution, before reducing this number to two. Scorpions and harvestmen have one pair of median eyes, as do the real spiders (where these eyes are called main eyes). In crabs, they are called nauplius eyes, and in insects, they are called frontal ocelli. Median eyes are altogether absent from centipedes and millipedes.

Mirror eye: A type of eye with light-reflecting structures for the gathering and focusing of light onto the retina. The reflective surfaces are either present in the entire background of the retina (tapetum lucidum), or in the background of the eye resembling a parabolic mirror, projecting light from the back onto the retina (as in the scallop species pecten). Reflective surfaces are also present in the crystal cone interior surfaces in the facet eyes of certain crab species.

Nauplius eye: See median eye.

Ocellus: A simple photo receptor with or without a lens, and with or without pigments.

Ommatidium: A unit of construction and of visual perception in the facet eyes of arthropods. In primordial arachnids (limulus), it is composed of a plano-convex corneal lens located above retinula cells. In crabs and insects, it is composed of a cornea, a four-part crystal cone and a rim of eight elongated retinula cells, containing rhabdoms of microvilli at their centers. The grouping of multiple ommatidia produces a facet eye.

Opsin: The protein component of a visual pigment, consisting of a protein and a chromophore – usually, either a 11-cis-retinal or a 11-cis-dehydroretinal.

Pax6: Belongs to the family of Pax genes, which partake in the development of eyes (and of other sensory organs) as so-called "Pax6 master control genes". Pax proteins (Pax being an abbreviation of "paired box" genes) are important in early embryonic development of animals, and affect the differentiation of tissue. They are also essential for the regeneration of lost limbs in species capable of such renewed growth. They are classified into numbered Pax gene groups. The Pax6 protein function is present in all animals – from coelenterates to humans – and is thus referred to as "highly conserved". For instance, the Pax6 gene of mice is able to trigger eye development in fruit lies (drosophila). What's more, the Pax6 genes of mice and humans are completely identical in structure.

Pit eye: A visual organ whose sensory epithelium lies within a pit-shaped indentation. Most eyes of this type possess a layer of pigment, and thus carry the name pigment cup ocellus.

Porphyropsin: The visual pigment in the cones of human eyes and of freshwater fish. It belongs to the retinal-2 type. Its name stems from its purple color. It absorbs light in the green part of the spectrum.

Pupil: The pupil is the aperture of the eye surrounded by the iris, through which light proceeds towards the retina.

Through a diminution (miosis) or an augmentation (mydriasis) of the pupil caused by its own muscles, the amount of inciding light is adjusted. While a wide-open pupil is always round, the shape of the narrow pupil differs from species to species.

Retina: In lens eyes, the retina is a layer of photo receptor cells consisting of different types of light sensory cells (rods and cones in the eyes of vertebrates). In facet eyes, these are called retinula cells. In each ommatidia, groups of eight such eyes form a rhabdom, and when put together, they produce a total retina – the sum of all light-sensory cells in an eye. In lens eyes, it forms a closed, planar epithelium. In facet eyes, groups of eight light-sensory cells are arranged in each ommatidium, which are then called retinula cells.

Retinula: See retina.

Rhabdom: The sum of eight receptor parts of the retinula cells in a single ommatidium. It is composed of finger-shaped protrusions of the cell membrane, in which the visual pigments are located.

Rhabdomer: The microvilli seam of a rhabdom's single retinula cell in the ommatidium of arthropods.

Rhodopsin: One of the visual pigments in the retina of vertebrate and invertebrate eyes, also called "visual purple" due to its color. In the rods of the human retina, rhodopsin is responsible for the differentiation of brightness and darkness.

Rod: A type of visual cell in the retina of vertebrates for the differentiation of lightness and darkness.

Selection: The difference in reproductive success of two individuals or populations of a single species due to differing genetic fitness.

Sexual selection: It rewards behavior or structures that increase the chances of successful reproduction for an individual. A frequent consequence of sexual selection is the development of spectacular feathers or combat structures (such as horns, claws, or jaws).

Sklera: The outer coating of the eyeball, reaching from the edge of the cornea towards the optical nerve. Together with the cornea, it forms the outer skin of the eye and is also responsible for its consistent shape.

Spherical lens: A type of lens that is found in many species of fish.

Stemma, stemmata, larval eye type: Occurs in the larvae of holometabolic insects, such as in butterfly caterpillars, who have 6-7 of such eyes on each side of their heads. They represent modified descendants of facet eyes.

Superposition eye: A type of facet eye where the ommatidia are not isolated by shield pigments, causing the light rays to incide simultaneously onto the rhabdomes of multiple adjacent ommatidia. They are usually found in nocturnal insects.

Tapetum, tapetum lucidum: A light-reflecting layer behind the retina in some nocturnal mammals (cats, dogs, and also horses). It can also be found in the auxiliary eyes of spiders.

Telescopic eye: The facet eye of crabs located on a long, flexible "stalk". Some species of flies and hammerhead sharks have non-flexible telescopic eyes.

Tetrachromatic: A type of eye with four different receptors in the retina.

Third eyelid: The *plica semilunaris conjunctivae* or membrana nicitans, is an additional fold of conjunctiva occurring in the nasal corner of many vertebrate eyes. It is transparent, and can be folded in front of the eye for safety. In humans, as in most apes, it is only present in a very rudimentary manner.

UV vision: The ability to see ultraviolet light present in some insects, birds, and fish. UV light is short-wavelength light with a wavelength below 400nm.

Literature selection

Eyes
G. Heldmeier, G. Neuweiler **Vergleichende Tierphysiologie** Bd. 1 Neuro- und Sinnesphysiologie (2003): Springer Verlag Berlin, 779 S
S. Ings **Das Auge. Meisterstück der Evolution** (2008): Hoffmann und Campe, 397 S
M. L. Land, D.-E. Nilsson **Animal eyes** (2012): Oxford University Press, 2nd ed., 271 S
I. R. Schwab **Evolution's Witness. How Eyes evolved.** (2012): Oxford University Press, 306 S
F. G. Barth, P. Giamperi-Deutsch, H.-D. Klein (eds) **Sensory Perception: Mind and Matter** (2012): Springer Verlag, Wien New York, 400 S
R. Wehner, W. Gehring **Zoologie** (2007): G. Thieme Verlag, Stuttgart, 954 S
D. E. Nilsson **Eye evolution: A question of genetic promiscuity** (2004): Curr. Opin.Neurobiol. 14 (4): 407-414

Evolution in general
F. J. Ayala **Die großen Fragen – Evolution** (2013): Springer Spektrum, 208 S
H. Burda, S. Begall, J. Zravý, D. Storch **Evolution: Ein Lese-Lehrbuch** (2009): Spektrum Akademischer Verlag, 493 S
D. J. Futuyma **Evolution: Das Original mit Übersetzungshilfen** (2007): Easy Reading Edition- Spektrum Verlag, 624 S
E. Mayr **Das ist Evolution** (2005): Goldmann Taschenbuch, 384 S
N. Shubin, S. Vogel **Der Fisch in uns: Eine Reise durch die 3,5 Milliarden Jahre alte Geschichte unseres Körpers** (2011): Fischer Taschenbuch Verlag, 281 S
V. Storch, U. Welsch, D. Arendt, T. Holstein **Evolutionsbiologie** (2010): Springer Verlag, 540 S
V. Storch, U. Welsch, M. Wink **Evolutionsbiologie** (2013): Springer Spektrum, 600 S
A. Wagner **Das Tier in Dir** (2013): Frederking & Thaler Verlag, 192 S
R. Dawkins **Geschichten vom Ursprung des Lebens: Eine Zeitreise auf Darwins Spuren** (2008): Ullstein Verlag, 928 S
Orig.: **The Ancestor's Tale: A Pilgrimage to the Dawn of Evolution** (2005): Houghton Mifflin

Index

360° perspective 62
aberration 121
Acilius 106
acommodation 76 126 129
acoustic reception 152
allele 12
ampullae of Lorenzini 83 92 154
Angler or Frogfish (Antennarius spec.) 188
Ant 22
antenna 65
antennapedia gene 147
apical 95
apposition eye 20 40 99
Arachnids 100 106
Arthropod 40
articulate animal 22
artificial selection 10
Asian Mantis (Hierodula patellifera) 66
astigmatism 118
auxiliary eyes 113
auxiliary retina 81
Axolotl (Ambystoma mexicanum) 43
axon 21 94
Baboon (Papio) 63
Beluga (Delphinapterus leucas) 139
bifocal lenses 106
binocular vision 65 67 83 96
bioluminescence 140
Blacktip Reef Shark (Carcharhinus melanopterus) 53
blind spot 95
blinking 203
Blowfish (Tetraodon spec.) 205
Blue-footed Booby (Sula nebouxii) 127 128
Blue Dot Ray (Taeniura lymma) 55 70 71
Blue Spotted Puffer (Canthigaster solandri) 189
body language 195
Butterfly Caterpillar 115
Butterflyfish (Chelmon rostratus) 132
calcite 104 121
Cambrian 109 147
Cambrian explosion 147
camouflage color 177
Carpus 149
cartilaginous fishes 54
cat eye 91 93
Cat Shark (Scyliorhinidae) 52
Caudate 42
Central Ommatidium 40
Cephalofoil 82
Cephalopods 94 109
Chafer (Chrysina gloriosa) 99
Chameleon (Chamaeleonidae) 76
Cheetah (Acinonyx jubatus) 89
Chelicerae 36 75 117
chemical evolution 10
chemical senses 153
chitin 167
chitin cornea 19
chromatophores 73
chromosome 11
cilia 104 148
ciliar muscle 19
Cleopatra Butterfly (gonepteryx cleopatra) 180
Cnidarians 18,112
color blindness 52 91 95
color cell 73
color perception 39 170
Common Cleaner Wrasse (Labroides dimidiatus) 171

Common Horse Fly (Haematopota pluvialis) 179
Common Newt (Triturus vulgaris) 42
Common Octopus (Octopus vulgaris) 72
Common Ostrich (Struthio camelus) 50
complex eye 18 148 179
Conehead Mantis (Empusa pennata) 67
connective tissue 93
contrast detection 39
contrast differentiation 93
convergent evolution 53
converging lens 80
Cooter (Trachemys scripta) 131
cornea 24 36 93 94 95 121 126 130 133 134 139
Cougar (Puma concolor) 88
countenance 184 195 196 206
Crayfish (Astacus astacus) 33 40
creation myth 10
Cretaceous 192
Cricket (Acheta domesticus) 118
Crocodilefish (Cymbacephalus beauforti) 177 201
crocodile tears 201
cryptochrome 155
crystals 121
Cucumber Green Spider (Araniella cucurbitina) 158
cuteness 198 204 206
cuticle 106 118
cuticula lens 179
cyclops eye 112
Daddy Longlegs (Pholcus phalangoides) 75
Damselfly (Zygoptera) 69
Darwin, Charles 6
Devonian 106
dichromatic vision 171
diopters 76 126 139
dioptric apparatus 106
DNA 146
Domestic Cat (Felis catus) 90
Dovetail (Papilio machaon) 114
Dragonfly 22
Drosophila 147
Dryas iulia (Heliconiidae) 181
Dugesia spec. 147
Earth's magnetic field 83 155
East Pacific red octopus (Octopus rubescens) 75
echolocation 67 127 153
Ectoderm 95
embryo 14 95 146
Emerald Tree Boa (Corallus caninus) 164
emotions 196
Emperor Nautilus (Nautilus pompilius) 108
Emu (Dromaius novaehollandiae) 51
epidermis 18 105
everse lens eye 51
EvoDevo 146
evolutionary biology 14
eye construction 166
eyelid 93 135 185 203
eye of the Hippopotamus (Hippopotamus amphibius) 137
eye rotation 35
Eyespot Puffer (Tetraodon biocellatus) 204
eye stripes 172
eye types 13 147
facet eye 27 28 32 35 36 37 39 40 58 62 65 67 69 100 106 110 112 115 118 120 121 127 179 208

facet sphere 112
Felidae 88
females of the Large Red Damselfly (Pyrrhosoma nymphula) 119
field of vision 62 85
Fire Salamander (Salamandra salamandra) 43
fish-eye lens 134
Flying Fox (Pteropus) 122
flying insects 22
focal length 100 126 132
focal point 104 121 128
focusing 91
Formicines (Formicinae) 58
fossil 109 121
Four-Eye Fish (Anableps anableps) 126,134
fovea 85
frontal eye 105
Frontal ocelli 106, 118
Gaboon Viper (Bitis gabonica) 203
Galapagos Sea Lion (Zalophus wollebaeki) 130
Gauß, Carl Friedrich 6
Gelada (Theropithecus gelada) 194
genetics 146
germ cell 142
Giant Squid 208
gills 42 111 174
glass pane 126
gonad 204
Gossamer-winged Butterfly (Chilades pandava) 187
Grass Snake (Natrix natrix) 44
Greater Horseshoe Bat (Rhinolophus ferrumequinum) 123
Great White Shark (Carcharodon carcharias) 92 93
Green Lacewing (Chrysopidae) 179
Green Tree Python (Morelia Viridis) 165
guanine crystals 93
Guitarfish (Rhinobatos spec.) 55
gustatory perception 152
Haeckel, Ernst 14
Haliotis spec. 104
Hammerhead Shark (Sphyrna tuburo) 81 82
Harvestmen (Opilio sp.) 74
Hawker (Aeshna spec.) 69
head rotation 29
heredity 146
Hermit Carb 37
hexagon 31 121
Hippopotamus (Hippopotamus amphibius) 136
holochroal eye 121
Holometabolic Insects 106
homeobox 147
homeotic genes 147
homology 148
Honeybee 22 99
horizon detection 118
Horned Frog (Ceratophrys spec.) 176
Horned Viper (Vipera ammodytes) 202
Horsefly (Tabanus bovinus) 179
Horseshoe Crab (Limulus) 36
Housefly (Musca domestica) 22 24
hox 104 146
humerus 21
Hummingbird (Trochilidae) 208
hygroreception 153
hyperacuity 75
image raising 24
Indonesian grasshopper 7
Indonesian shrub frog 46
infrared 99 154 164
interference filter 179

inverse retina 100
iris 93 184
Jacobson's organ 162 164
Japanese Oak Silkmoth (Antheraea yamamai) 115
Japanese Robberfly (Dioctria nakanense) 31
jaws 45
Katydid (Tettigonia viridissima) 30
Komodo Dragon 163
Kraken (Octopoda) 73 96
labial pits in Boas 164
Lamarck, Jean-Baptiste 10
larval eyes 106 115
lateral eyes 100
law of refraction 128
lens 93 95 101 106 115 118 126 127 129 132 139 143
lens camera eyes 94
lens eyes 18 27 29 76 100 148 208
lens system 139
Leonardo da Vinci 6
light-sensory organs 118
light efficacy 80
Limulidae 36 121
Lionfish (Pterois volitans) 175
locus 12
Longhorn Cowfish (Lactoria cornuta) 174
Lynx 89
macromolecules 149
magnetic fields 152
magnetite 155
Mantis (Mantis religiosa) 64 65 118
Mantis Shrimp (Odontodactylus spec.) 39 64 98 99
Marbled Rock Crab (Pachygrapsus spec.) 38
mathematics 6
Maybug (Melolontha spec.) 157
mechanoreception 117 152 153 159
median eyes 100 106 118
meiosis 11
melanin 184
membrane 152 159
Mexican Dwarf Orange Crayfish (Cambarellus patzcuarensis) 40
Microbat (Microchiroptera) 122
microorganisms 112 146
microvilli 95 148
Migratory Locust (Anacridium) 119
Millipede (Alcimobolus domingensis) 110
Mimic octopus (Thaumoctopus mimicus) 75
Mindoro Crocodile (Crocodylus mindorensis) 201
miniature eyes 112
mirror optics 40
mirror reflection 80
mitosis 11
Mole Cricket (Gryllotalpa gryllotalpa) 32
molecular cascade 170
molecular genetics 137
Mole Salamanders (Ambystomatidae) 43
monofocal lenses 88
Moray (Muraenidae) 45 48
Morpho (Morpho peleides) 186
Moth 20
Mudskipper (Periophthalmus) 134 135
multifocal lens 91 130
muscle contraction 129
mutation 142 146 185
natural selection 10 146

Nautilus 104 109
Neotrygon Kuhlii 54
neuronal signals 154
neurotoxin 204
Nile crocodile (Crocodylus niloticus) 200
ocelli 73
Octopuses (Coleoidea) 94
oculesics 184
olfactory epithemlium 153
olfactory reception 153
olfactory sensor 157
Olm (Proteus anguinus) 43
ommatidia 19 27 58 69 80 98 100 111 112 115 179 180
ontogenesis 14
opisthosoma 36 75
opsin 105 148 170
optical axis 99
optical nerve 62 93
optic chiasm 62
oral jaw 48
Orang-Utan (Pongo pygmaeus) 193 196
Orange-striped Watchman Goby (Amblyeleotris randalli) 189
Ordovicium 109
Oriental Garden Lizard (Calotes versicolor) 57
Osteichthyes 48
oval pupil 62
ovum 12 146
paleontology 10
palpebra 57
Pax 104 146 209
Peacock Mantis Shrimp (Odontodactylus scyllarus) 65
Phacopida Trilobites 121
phacops lens 121
photon 62 170
photopigments 95
photoreceptor 13 104 105 148 152
phylogeny 14
physiognomy 184
pigment eyes 113
pigments 170
Pilot Whale (Globicephala melas) 137
pinhole camera eye 18,109 148
Pipistrel (Pipistrellus pipistrellus) 160
pit eye 104 148
pit organ 164
pixel 22
Plaice (Pleuronectes platessa) 71
point eye 73
polarisation 65 99
pollen 167
Polychaetes 18
polyp 73
Porcupinefish (Diodon holocth.) 205
Porcupine Puffer (Diodon hystrix) 204
porphyropsin 42
prism 120
proof 6
prosoma 36 75
protective function 114
protein 147 148
Pseudanthias squamipinnis 172
pseudo-eyes 65
pseudo-iris 177
pseudo-pupils 180
Pug (Canis lupus familiaris) 199
pupal stage 115
pupil 63 73 88 91 93 96 128 130 140
Pygmy Damselfly (Nehalennia speciosa) 68

reaction time 29
receptor 85 95 133
receptor protein 13
Red Rock Crab (Grapsus grapsus) 38
reflection 91 120
refraction 24 126 128 133 139
refractive lens 18
refresh rate 85
regulatory cascade 147
residual light amplifier 90
resolution 84
retina 19 46 55 62 80 85 93 94 95 96 101 104 106 118 120 126 130 133 148 170
retinomotoric change 133
retinula 19 40 112
rhabdom 19 40 69 148
rhinoptera bonasus 54
rhodopsin 42 104 170
Ring-tailed Lemur (Lemur catta) 185
ring muscle 19
Robber Fly (Dioctria linearis) 27
rod 81 93 127 170
rod cells 133
rod pigment 170
rod type 53 130
Rose Chafer (Cetonia aurata) 166
Rough Woodlouse (Porcellio scaber) 111
Sandtiger Shark (Carcharias taurus) 92
Satyrines 187
Scallop (Pecten) 81
Scalloped Bonnethead (Sphyrna corona) 82
schizochroal 121
Sea Goldi (Pseudanthias squamipinnis) 171
Seal (Phoca situlina) 139
Sea wasp 113
selection 11 13 143 146
sense of touch 158
sense of vibration 152 158 159
sensory body (rhopalium) 113
sensory cells 62 95
sensory organ 12
sensory world 152
Sepia (Sepia officinalis) 97
sexuality 11
sexual pheromonoes 153
sexual selection 179
Sharksucker (Echeneis spec., Echeneidae) 205
Sheep (Ovis spec.) 167
short-sightedness 134
Siberian Tiger (Panthera tigris altaica) 184
Silurian 100
Snow Owl (Bubo scandiacus) 63
spatial vision 30 62 96 99 101
sperm 12
Spotted Congo puffer (Tetraodon schoutedeni) 133
squaring the circle 6
Squid (Loligo spec.) 96
squinting 35
stemmata 106 115
superposition eyes 21 80 403
synapses 153
Tapetum Lucidum 81 93 100 130
Tarsier (Tarsius tarsier, Tarsiiformes) 29
tear fluid 131 185
Telescope eye goldfish (Carassius gibelio f. auratus) 143
telescope eyes 82
telson 65
tetrachromatic 100

tetraradial 113
thermonectes 106
thermoreceptors 153
third eyelid 45 56 57 86 93 128
thorax 147
Tiger (Panthera tigris) 89
Tiger Fly (Coenosia tigrina, Muscidae) 22
Tiger Mosquito (Stegomya albopicta) 156
tongue 162
tracheal system 110
transcription factor 147
tree from (hyla arborea) 47
Tree Shrew (Tupaia) 192
Trilobite 18 121
tubular eyes 81
turbulence 117
tympanal organ 152
ultrasound 123 160
ultraviolet 99
UV light 85
Vapourer (Orgyia antiqua) 114
variation 146
Veiled Chameleon (Chamaeleo calyptratus) 76
venom 162
ventral eye 127
vibration 117
viewing angle 62
viewing direction 35
visual acuity 88
visual cell type 100
vitreous body 93
Wandering Spider (Cupiennius getazi) 159
Water Flea (daphnia pulex) 112
Water Frog 44
Weever (Trachinidae) 70
Whale Shark (Rhincodon typus) 141
Whirligig Beetle (Gyrinidae) 127 152
Whitetip Reef Shark (Triaenodon obesus) 53
Woodlouse 111
Yellow-legged Robberfly (Dioctria linearis, Raubfliegen) 28
Yellow Sac Spider (Cheiracanthium) 116
young medusae of the Box Jellyfish (Tripedalia cystophora) 113
Zebra Spider (Salticus scenicus, Salticidae) 23
zonule fibers 19

To be continued …

Printed by Printforce, the Netherlands